Design for Configuration

Springer-Verlag Berlin Heidelberg GmbH

Asko Riitahuhta and Antti Pulkkinen (Eds.)

Design for Configuration

A Debate based on the 5th WDK
Workshop on Product Structuring

 Springer

Professor Dr. Asko Riitahuhta
Laboratory of Machine Design
P.O. Box 589
33101 Tampere
Finland
email: asko.riitahuhta@tut.fi

Antti Pulkkinen
Laboratory of Machine Design
P.O. Box 589
33101 Tampere
Finland
email: antti.pulkkinen@tut.fi

With 104 Figures and 12 Tables

ISBN 978-3-642-63211-2

Cataloging-in-Publication Data applied for

Design for configuration :
a debate based on the 5th WDK workshop on Product Structuring /
Asko Riitahuhta and Antti Pulkkinen (ed.). –

ISBN 978-3-642-63211-2   ISBN 978-3-642-56905-0 (eBook)
DOI 10.1007/978-3-642-56905-0

http://www.springer.de

© Springer-Verlag Berlin Heidelberg 2001
Originally published by Springer-Verlag Berlin Heidelberg New York in 2001
Softcover reprint of the hardcover 1st edition 2001

Cover design: medio Technologies AG
Typesetting: Camera ready copy by author

SPIN: 10774172   62/3020xv-5 4 3 2 1 0 – printed on acid-free paper

Preface

Product Structuring for Design Reuse, Product Platform systems, Configuration and Modularisation is gaining growing attention in global operating companies as means to enhance productivity in manufacturing of customised product variants. In that situation it was a great opportunity to get the 5th Product Structuring Workshop to Tampere. The special theme of the workshop was chosen Design for Configuration.

We see that Knowledge Systematisation is an important research area for Product Structuring. We made a decision to apply Knowledge Systematisation for the workshop itself: we used the small group working method-Double Team Method- in workshop discussions for creating adding values based on presented papers, and after workshop we have made a synthesis for crystallizing results from papers and group works.

On the behalf of organisers I thank for the enthusiastic participation on the all stages of workshop and Finnish Academy and Tampere Centre of Expertise for funding the workshop and publication of this book. I am sure that all participants would like to join to my many thanks for the workshop secretary, Ms Jenni Kauppila for her efficient organisation work. After workshop she has made excellent work as our editorial secretary.

We would like to give special merit to all the participants of the workshop for good discussion and ideas presented in the group work sessions. The participants were: Aasland Knut, Blackenfelt Michael, Hassinen Markus, Holmquist Tobias, Jensen Thomas, Jokioinen Ilkka, Juuti Tero, Karlsson Jan, Kostamo Juha, Kwok Sonny, Lange Mark, Larsson Leif, Larsson Tobias, Lehtonen Timo, McKay Alison, Mortensen Nils-Henrik, Oosterman Bas, Persson Magnus, Pulkkinen Antti, Riitahuhta Asko, Saukkosaari Tapio, Smith Joanne, Stake Roger, Strömbäck Kim, Suistoranta Seppo, Sören Andersson, Tichem Marcel, Tiihonen Juha, Vainio-Mattila Markus and Åslund Johan

Prof. Asko Riitahuhta   Tampere, Spring 2001

The co-organisers of the workshop:
- Institute of Machine Design, Tampere University of Technology, Finland
- Laboratory for Production Engineering and Industrial Organisation, Delft University of Technology, The Nederlands
- Department of Control and Engineering Design, Technical University of Denmark
- CAD Centre, University of Strathclyde, United Kingdom.

Contents

4 Part III: Metrics and Methods for Modularity and Configurability

5 Part IV: Supporting Modeling and IT-Tools

Introduction

Views and Experiences of Configuration Management

Asko Riitahuhta

Laboratory of Machine Design
Tampere University of Technology
P.O.Box 589
33101 Tampere, Finland

1 Introduction

Product configuration is an issue related to evolution of software tools, business paradigms and product structuring. Experiences on developing software solutions for configuration purposes date back to 1980's, when first configurators were introduced. Since then a lot of experience from implementing configurators has been gained.

Our insight into configuration comes from three different sources. First, a decade ago engineering companies developed Engineering Configurator Systems. Second, Sales Configurators and e-business have been a commercial success; for that we have experiences from a Finnish Technology Development programme RAPID. Third, Knowledge Systematisation is the important prerequisite for succesful Configuration and Modularisation is the most important means to achieve it.

Profitability has been driving companies for global mass production. However, globally operating companies have realised that locally customised products and services are necessary for success in global business environment. This goes against the original mass production paradigm. It creates the need for a capability to understand customers better. It is essential for the sales organisations to pay as much attention as possible to the customers' needs and demands. The new paradigm, *Mass Customisation* [1], strives for the management of product variants in order to meet local customer's demands.

Other driving forces for the search of new post mass-production -paradigm are:
1) The need to add value adding features and services to the products in order to differentiate from competitors
2) Growing environmental consciousness that calls for product maintenance, sub-unit recycling and upgrading of environmental strategies
3) The trend that big companies concentrate in their core competence areas and form supply chains with their subcontractors.

In addition to the mass customisation-paradigm there are several other post mass-production-paradigms, for instance agile manufacturing, precision strategy, and inverse manufacturing, which has lately turned into one of the key research areas in Japan.

Joseph Pine II defines mass customisation as the production and distribution of customised goods and services on a mass basis. Brian Maskell [2] defines agile manufacturing as the ability to use flexible and diverse production environment to create quickly products that meet or exceed customer's expectations for customisation, variety, cost and quality. Configuration management and modularisation are rudiment for moving towards the post mass production paradigms [2].

We call the management of product and service variants as Configuration Management. The creation of a particular variant, a product for instance, is a configuration task. Software tools have been developed to perform the configuration task.

Successful Configuration management contributes to satisfying customers and getting profitability for the company. If Configuration management is not done correctly, unplanned extra work in all functional organizations of a company usually emerge. The result is extremely high profit losses.

Configuration Management is gaining importance when initiatives such as Global Engineering Networking, GEN are being realised [3]. GEN supports engineers and companies by being an electronic marketplace. It is a medium for virtual enterprise, where tele co-operation is a key-issue. The crucial elements in Configuration are: precise fitting of the products to the needs of the customers, creation of a competing edge through unique, requested features and dynamic, strongly competitive development of the markets.

Product family development consists of different degrees and types of systematisation. Configuration management demands the systematisation and structuring of the product assortment. Development of a product family must also contribute to relevant product range for the market and rational manufacturing. At least the formulation of rules for the assortment is needed. We see this kind of production preparation as modularisation in a broad sense. With this interpretation we look closer upon the possibilities and positive effects of modularisation as a means for configuration.

The rest of this paper is structured as follows. We define Configuration Management in the second section. Then both engineering and sales configuration are described with cases. Knowledge management research is studied in section five and it is followed by a section on the relation between modularisation and configuration. Seventh section describes the development stages of Modular Engineering Systems. In it we present our experience of a research project on configuration of a virtual organisations. In section eight we describe Dynamic Modularisation paradigm, which aims for avoiding drawbacks of modularisation and for utilising engineering networking. Section nine is about evaluation and configurability. The configuration management principles are outlined in section ten, which is followed by conclusions.

2 Configuration Management

2.1
Definitions

We adapt the following definition for the configuration from the GNOSIS-project of the global IMS-program [4]:

Configuration Management Systems comprise the processes of creation, modification and control of products with many variants, together with all the technical and technological information, which represent them throughout their life cycle. They also take the environmental and organizational context of the product's life cycle into account, which is important for the continuous product improvement process.

A much-referenced definition of Mittal 1989 [5] of the configuration task tells the following:
Given:

> *(A) A fixed, predefined set of components, where a component is described by a set of properties, ports for connecting it to other components, constraints at each port that describe the components that can be connected at that port, and other structural constraints, (B) some description of the desired configuration; and (C) possibly some criteria for making optimal selections.*

Build:

> *(B) One or more configurations that satisfy all the requirements, where a configuration is a set of components and a description of the connections between the components in the set, or, detect inconsistencies in the requirements.*

A *functional architecture* specifies a functional decomposition of the artifacts and constraints on their composition. It is shown that introducing a functional architecture can reduce the complexity of the general definition of configuration.

In the following we will describe the relation between product structuring and configuration. We call it Design for Configuration, DFC.

3 Engineering Configuration Systems

From the beginning of 1980's there has been efforts to develop Expert Systems (sometimes called Knowledge-Based Engineering Systems) for capturing product knowledge and creating product variants, i.e. configurations. These efforts have been taken both in industry and academia.

For instance, Digital Equipment developed software tools, called XCON and XSEL, for configuring VAX computers and accessories based on a customer specification. Bechtel, Tampella, Ansaldo and other process plants manufacturers developed configurators for conceptual design.

The early configurator projects required that all knowledge was gathered from the developing company itself, because no commercial knowledge bases existed. If knowledge changed rapidly, a system was difficult to keep up-to-date. This required strong management belief and commitment inside the companies to develop and maintain expert systems.

There was also active university research going on. Its goal was to develop knowledge bases more useable, maintainable, and understandable for designers and other users. One of these projects was SHARE led by Prof. L. Leifer at Stanford University [6]. SHARE emphasised teamworking between multi-disciplinary organisations, supported by networking techniques and IT-tools. The objective of SHARE was "...*to provide the enabling technology that will support design engineers by allowing them to access helpful information over the network not typically available to a single user environment*". Another approach was taken by Cutcosky et al. (1993) for the PACT project at Stanford.

Also in Europe there existed several Knowledge Systematisation projects. For instance a model based Knowledge Based System for a product configuration was developed at the Tampere University of Technology. The project was called Automatic Component Selection (ACS).

3.1
Case 1: Automatic Component Selection (ACS), Configuration process in Engineering Configurator

For the verification of the development of ACS we had industrial examples; Make-To-Order products like cranes, diesel power plants and conventional power plants. Product architecture can be configured by using rules, parametric design and component libraries, in that kind of products. The product architecture is the most important knowledge sharing aid when, for example, a process plant is configured. In the development of configurator we went through the following steps [7]:

1. We started combing Knowledge-Based Engineering System with CAD-system for configuring a product structure, architecture and layout. Later on we combined Knowledge Base System and CAD with a relational database.

2. In large system products, the component selection is the key issue. In the development of a prototype for a large component selection system we realised that the combination is not efficient enough. Therefore, we started to use Object-Oriented Data Base Management System, OODBMS. We created workflow for product configuration by utilising KBS, CAD, Documentation program and OODBMS, Fig. 1.

The configuration process used in the Automatic Component Selection, ACS project, starts with the customer's technical specifications. Usually these come from the sales persons. Creation of the product structure, i.e. the configuration work is done by selecting a suitable product family description from the database or by using the configuration rules from knowledge base. The configuration rules can include also selection of product family description. A product family description does not have to describe the whole product but a smaller part, or a subassembly, of a product. The system works even if the configuration model is not complete, because user can complement it with his own knowledge.

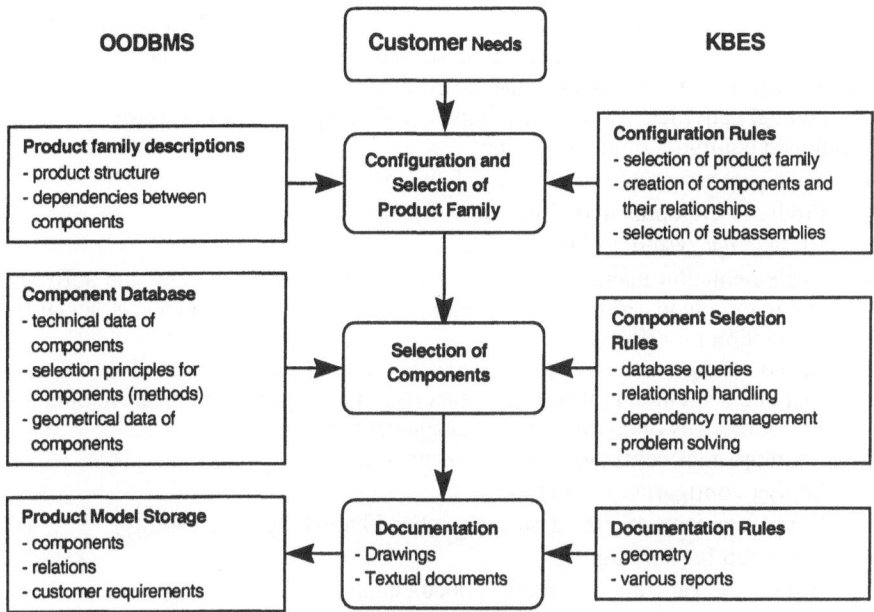

Fig. 1. Workflow of a product Configuration [7]

The product structure includes various relations between assemblies and components of the product. The most obvious are part-of relations between an assembly and its components. Other relations are, for example in diesel power plant, *fuel_out*, *power_chain*, *geometry*, and *identity* relations. These relations are created when the components are instantiated in the product model. In the product model, the relationships often form a chain or a net between the components. When for example a pump has to be selected, the required capacity can be calculated using the *fuel_out* relationships. All the components, and their pressure drops, in a chain between the pump and an engine using the fuel can be taken into account in the calculations. Geometrical relationships are used when the Process and Instrumentation (PI) diagram is generated. The geometrical relationships include information about how some components are located and connected to each other in a drawing.

After having defined the product structure, components are selected. This is done by sending a simple database query to the Object-Oriented Database Management System (OODBMS). The query includes the name of the class of the component to be selected, the name of the component, selection method, and arguments to that method. The arguments are collected from attributes of the component itself or from other related components' attributes. A method in the

OODBMS can include more or less complicated calculations, references to other classes in the database, and other criteria. The OODBMS sends back all technically acceptable components for the user to make a selection, or a single component, which is selected by using some optimization criteria, such as price or size.

Once the configuration and component selection are completed, the system creates a PI diagram as a document of the design. The design can also be documented with textual reports and stored in a database.

Essential activities in a full scale implementation project of the above mentioned configurator are following:

- Main processes and sub-processes have to be defined
- Product development, optimization of the product life cycle
- Knowledge systemisation
- Implementation manager a team from manufacturing company and software company (supporting research team from a university gives a scientific foundation for efforts)

Developed configurator can deliver the first results in 7 months, it will be capable in a year and a strength tool within 2 years from the beginning of the development.

General benefits of configuration management are:
More homogenous configurations for all customers

- Shorter configuration lead time
- Product variants are based on systematized knowledge (traceability)
- Less need for iteration
- Automated generation of product documents
- Product model available for maintenance

The system presented in Fig. 2 did work well and it fulfilled its task in proving the feasibility of the used techniques. However, during the development of the system the following weaknesses were noted:

- The design rules tended to become large and they were hard to understand especially for people who were not familiar with the LISP language.
- A change in the user specification usually means that a major part of the product model is removed and recreated, even though the new result would be (almost) identical to the previous one. To overcome this, the rules would have had to be much more complicated in order to take in account all possible changes.
- Manual changes in the design could be done, but the system did not have to check if the changes were consistent with the design rules.

These issues gave the motivation to develop the constraint language (CLAN) to check configuration under the guidance of a designer. The presented project was part of an global research programmme, Intelligent Manufacturing Systems, IMS. Industrial and academic partners from Europe, Japan, Canada and USA co-operate in several different projects in the intelligent manufacturing research field. One of IMS projects is GNOSIS. Its focus has been on how to systematize knowledge so that a product configuration is able to fulfill all highly evaluated demands of the human, society and environment.

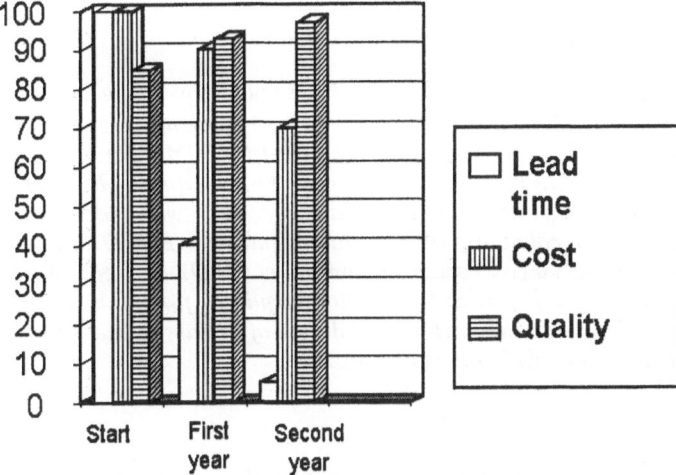

Fig. 2. Results of the implementation of a configurator. The estimation is based on the experiences from one large implementation and two demo projects in multidiscipline products.

4 Sales Configurators

4.1
Commercial Software

The development of Engineering configurators did not continue as it was estimated because of the above mentioned difficulties. However, at same time sales configurator systems were developed and used. Knowledge capturing for those systems was much easier, systems were cheaper and results were acheived faster than with engineering configurators. Motivation for using sales configurators was very obvious. *Misconfigured orders can cost as much as 2% of corporate revenue* according to Trilogy Company. In their web-site they present following [8].

The challenges of selling customizable goods and services online: Customizable goods and services-computers, automobiles, and equipment-generally allow customers to select optional parts, accessories, or functionality to suit their needs. A customer can easily be overwhelmed with the variety and complexity of a configurable product. If offerings cannot be tailored to customer's specific needs, or if the buying process is too difficult, the customer will go elsewhere. In addition, order accuracy is essential in ensuring customer's satisfaction. For example, if a product is incorrectly customized with parts that don't work together, the customer will be frustrated and the enterprise will be burdened with the return of the good. Many enterprises, therefore, only offer a subset of their full products and services online, thus limiting their ability to compete.

Sales configurators have been grown or they have been bought by bigger software companies. Today sales configurators might be a part of integrated, web based software. One example is RHYTHM solutions [9]: *i2's RHYTHM solutions offer the intelligent answer for decision-making across the enterprise. RHYTHM software optimizes and integrates key business processes, while delivering intelligent eBusiness through collaboration with trading partners. RHYTHM offers a complete solution for Business Process Optimization (BPO) by offering the optimization, integration, and forward visibility required for high-velocity business. The RHYTHM solution has delivered billions of dollars in measurable value for major companies in a wide range of industries.*

4.2
Case 2: Improving Product Development Productivity

Developing modularized project delivery products to satisfy individual customers' needs, and the use of an integration software [10] The case company (Berendsen PMC Oy Ab) aimed at Product management with sales configuration. According to the Berendsen's Divisional Manager "…the company has to offer a wide range of individualized products, which are based on a varied product structure and a product configurator." Furthermore, this has to based on " …an individualized but explicit property-based product range". The company also aimed at independent product management by company's own external sales network. The company expected results like high customer response in sales process"…customer contacts should lead to deliveries as often as possible".

4.2.1
Focus on Product Projects

The primary object was to develop a mass customized hydraulic actuator. The work was based on an existing cylinder series, which accounts for more than half of the company's sales volume and for which a property-based sales configurator and a production-based product configurator were to be drawn up.

Adopting a property-based approach proved difficult, even for sales staff. It seems to be easier to think on the basis of the product's traditional technological content than on the basis of the additional benefits accruing to the customer and the properties this approach requires. However, during the project, more and more of those involved embraced the new thinking.

Though simple-looking, the hydraulic cylinder proved difficult to modularise. This was particularly true in the case of industrial applications because modularisation led to a surprisingly large number of variants. The fact that the stroke or swept length of the cylinder was freely variable was also problematic, as it made cost accounting difficult. On the other hand, this forced the company to think more carefully about the number of interchangeable parts in each production series. Their number had to be maximised, as required by the fundamental principles of modularisation.

4.2.2
Making Company Processes Compatible with the New Operational Requirements

An essential part of the development work described here was to make the company's processes compatible with the project targets. The company's information system is crucial in this respect. Originally the configurator was to be integrated in the information system of the unit in question and external sales units and customers were to be provided with 'stand-alone' configurators, which would be updated by the unit. It became clear during the project, however, that the unit's information system should be integrated into the owner's global sales system through which all sales (including those of the unit in question) are processed. The fact that the supplier of the configuration system is now part of an American company has made the network solution a viable alternative. In fact, rapid developments in information technology have probably made it the only feasible option. Unfortunately, the prices have also reached levels that Finnish SMEs are not used to. Thus, it has not been possible to introduce the originally-conceived system on schedule. The question of system integration, the policy concerning e-commerce and the way customers can shop via Internet must first be dealt with.

4.2.3
Benefits

The original targets have been met in that the customers are now guaranteed products tailored to their needs. Cost management is on a sounder basis and geared to exceptional situations while product development is more flexible. Moreover, life cycles are well-managed and fewer components are kept in stock.
The company has seen its competitiveness improve, and the availability of the products is on a more global basis.

Most of the strategic targets have been reached. As product management has improved, the operations are more effective. Moreover, operational respon-sibilities have become clearer and delivery times can now be optimized for each customer. Product development has also become increasingly property-based and resources are now being transferred to product development and product/customer management. Finally, the process involving orders taking and deliveries has become leaner.

4.2.4
Required Fundamental Actions

No development project is an isolated effort. The company would have achieved some of the improvements described here on its own, although not all aims would have become reality without the project with a consulting company. None of the

targets would have been met, however, without the adoption of the following basic
principles:
- Company strategy and product/sales policy must be outlined and appropriate
  choices made.
- Company culture must be made customer-oriented so that it lends itself to
  teamwork
- Sufficient attention must be paid to the modernization and smooth functioning
  of the order/delivery process.

It has also become clear that there is still a lot to be done in improving product
management.

5 Knowledge Systematisation

The use of configurators has not grown as fast as was expected a decade ago. The
situation is the same in Product Data Management Systems. One reason is that a
product structuring was based on assembly and parts structures. A hierarchy
coming from manufacturing was used as a product structure. This caused that
product knowledge and information were difficult to search, update, and reuse.
Systems did not support knowledge outside of a company. Even though sales
configurators promised good results some companies with many product options
did not succeed in organising the needed information to their sales configurators.
This situation led to the rise of knowledge systematisation into an important
research area in the 1990's.

Knowledge systematisation was one major research project of the GNOSIS
project in Intelligent Manufacturing Systems-research programme. One of leading
researcher of GNOSIS, Prof. Tomiyama from Tokyo University of Technology
emphasises the importance of knowledge systemisation as follows [11]:

*"...we must focus more on knowledge as a source added value. By intensively
using product life cycle knowledge in an integrated manner, we can generate
more added value including innovation, longer life, higher reliability and
robustness, more flexibility, and cheaper life cycle cost. To this end, we propose
"knowledge intensive engineering" aiming at generation of more added-value
through intensive and integrated use of knowledge, facilitating mutual
communication among life cycle knowledge agents."*

In Europe the development of product catalogues and meta catalogues for
managing them have been the interest area in GEN initiative.

Knowledge systematisation research is fundamental when configuration
systems are developed. WDK design research community and Tokyo University
Knowledge sytematisation group suggest the use of abstract structures in product
knowledge structuring, such as function and process structures.

We have chosen to use design process which follows patterns as described by
Andreasen [13], Hubka [14] or Riitahuhta and Andreasen [15], see Fig. 3. The
pattern is an alternation between process, organ and building structure
considerations.

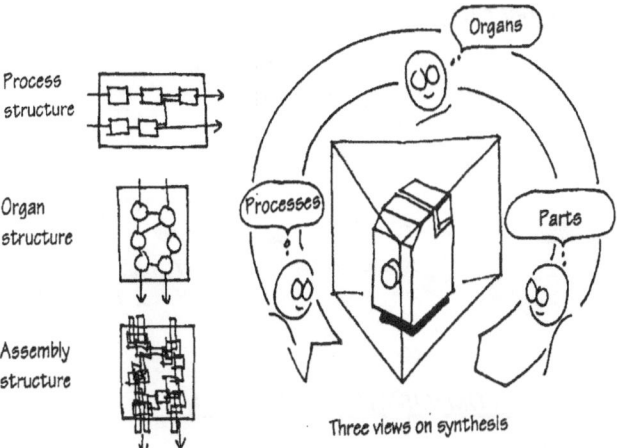

Fig. 3. Design seen as an alternation between process, organ and building considerations. [15]

Process considerations focus on products's use phase while organ conside-rations focus on the realisation of the required functions. The building structure considerations concern the machine parts and their assembly related interplay. The designer may choose only to design "concretely" in the building structure domain or he/she may alternate between more domains.

The result of the design work is a building structure (German: Baustruktur), i.e. a specification of the parts and their assembly. From this structure we may read both processes and organs, but normally they do not occur as distinct elements with simple interfaces.

In an earlier paper [Andreasen 1996] it was shown how one might superimpose the building structure of a product with structural compromises, which for instance makes the product easy to assemble. Such a restructuring, which may lead to a modular structure, is an important element in different types of DFX, for instance Design for Manufacture, Design for Assembly, Design for Quality and Design for Environment [17].

6 Configuration and Modularisation

How is it possible to combine configuration management and maintain the flexibility to apply revolutionary ideas of a customer? From the presented cases and other experiences with projecting business and Make-To-Order products we have composed the working model where customer demands and configuration management have been combined (Fig. 4).

As shown in Fig. 4 modularisation is knowledge-intensive process. Modularisa-tion comprises knowledge from different sources. We are developing the method-

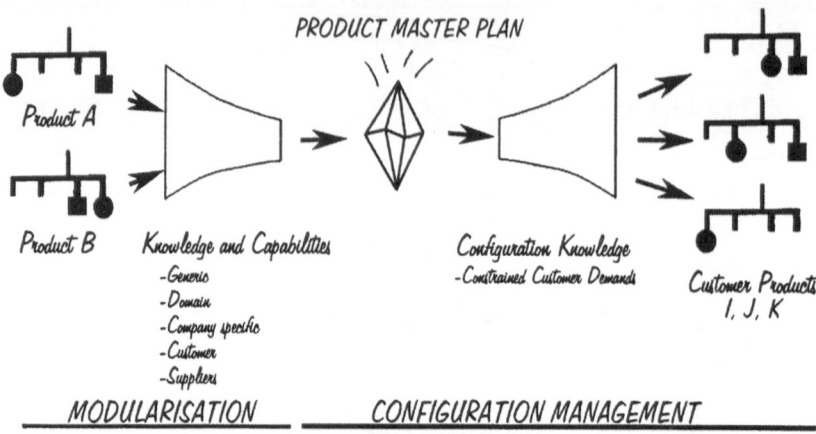

Fig. 4. Product Variants through Configuration Management

base to improve this process. Our aim is to combine the concept creation by integrating customers into the modularisation process. According to our insight it is possible to select the reference group from customers and customer's stakeholders to find ambiguous goals. We found modularisation on the company's existing products and its result is presented in Product Master Plan. The Product Master Plan is related to the calendar time, the new versions of it will have a certain period of validity. It does not pick up properties unmanageably from customer specific product deliveries.

Modularisation is the key issue in Configuration Management, but knowledge acquisition and sharing are also important. In the future it is a necessity to assure configurability at an early stage of product development. Usually it is not possible to cope with a closed configuration system, because configuration management has to cover whole life cycle of the product master plan.

For these reasons we suggest that the paradigm "Dynamic Modularisation" (see section 8) should be used in configuration management. Other conditions for the development are that the Configuration Management System should preferably include all structures of the Domain theory, because the organ structure offers a possibility to manage some aspects of the product in long run and the transformation process structure is important for sales.

7 Development Stages of ME Systems

A ME System evolves through a sequence of stages, which are presented below. We might not find clear-cut examples of enterprises that would have evolved through these steps, but we can find some examples, where current state and organisation are in accordance with the stage and where endeavours are to reach the next stage. Each stage contains it's network of involved internal functional areas and external stakeholders and thereby a set of corresponding metrics. Each stage may be seen as an enhanced instrumentation of business activities in each functional areas and an enhanced co-ordination and integration in general.

We have chosen the names and main characteristics of the five stages of development. Note how the range of the organisational structure and the importance of the modularisation are growing through the stages [15]. We use the abbreviations: PD/PP=Product Development/Product Planning, Ps=Products, Pr=Production, PL=Product Life aspects, KM=Knowledge Management aspects.

The five stages of development are:

1. *Manufacturing of independent products*
PD/PP: Independent development of each product, each product obtains its completing edge based upon its own functionality and qualities
Ps: No structured kinship between the products
Pr: No structured concern from design to production (no DFM effort)
PL: Product life aspects not utilised for competition
KM: No organised or structured knowledge handling

2. *Product families as market strategy*
PD/PP: Co-ordinated development of products into product families
Ps/Pr/PL/KM: As stage 1.

3. *Architecture related manufacturing*
PD/PP: Co-ordinated development of products into product families
Ps: Architectural commonality of a product family's products
Pr: Production planned for unit or module production
PL: Product life aspect to a certain degree utilised for competition
KM: Knowledge-management to a certain degree in accordance to product architecture

4. *Configuration-oriented manufacturing*
PD/PP: The product development is controlled and managed based upon family architectures, platform definitions and market needs
Ps: As 3
Pr: Production and suppliers productions are unit oriented
PL: Product Development is product life oriented
KM: Documentation and Knowledge Engineering structured like the family architectures.
Pr: Production and suppliers productions are unit oriented
PL: Product Development is product life oriented
KM: Documentation and Knowledge Engineering structured like the family architectures.

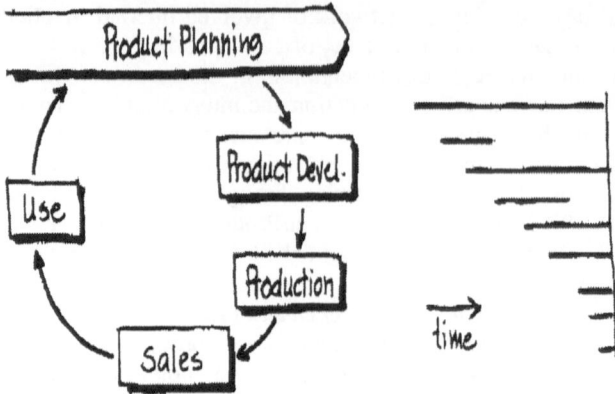

Fig. 5. Manufacturing of independent products

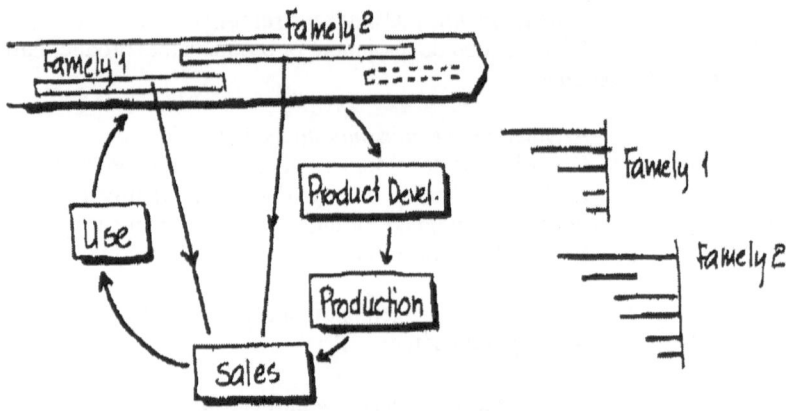

Fig. 6. Product families as market strategy

5.   *Dynamic modularisation approach*
        PD/PP: New product development based upon creating or purchasing
                        new units or modules, in accordance to a dynamic platform
                        and dynamic market demands
        Ps: Flexible architectures accepting new modules, deleting old

Pr:	Units and modules from a mass customisation production philosophy, also at suppliers
PL:	Product life fit and qualities utilised for competition
KM:	Dynamic innovation of module and architecture related knowledge. Knowledge Empowerment as a strategic element.

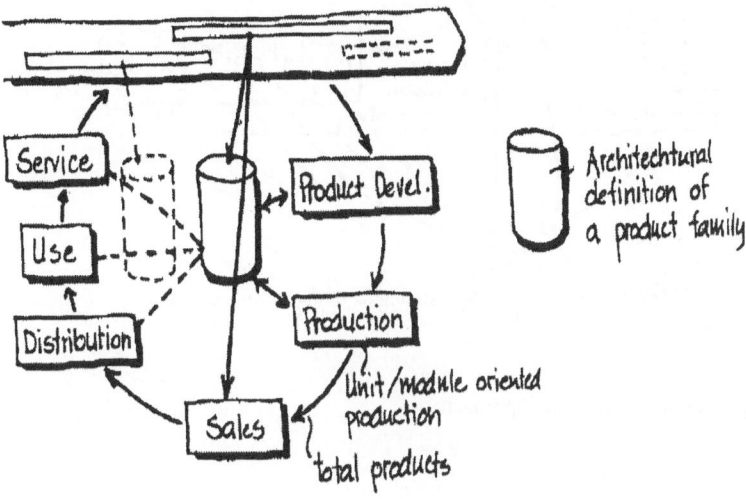

Fig. 7. Architecture related manufacturing

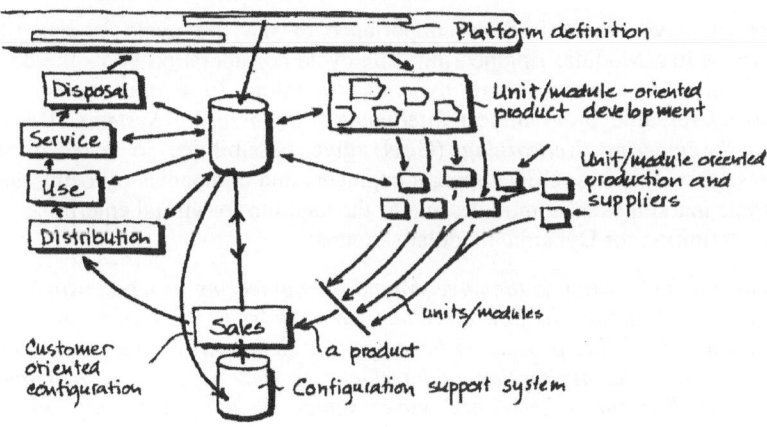

Fig. 8. Configuration-oriented manufacturing

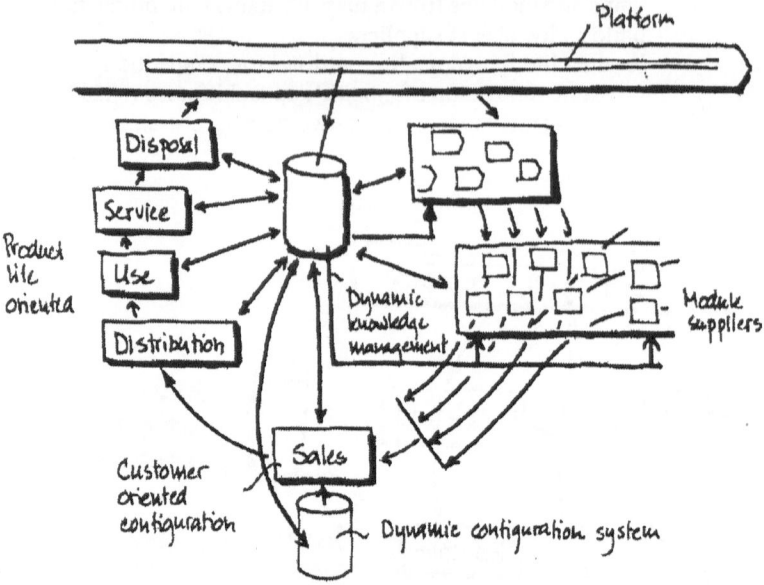

Fig. 9. Dynamic modularisation approach

8 Dynamic Modularisation (DYMO)

8.1
DYMO Definition

Above we have emphasised the importance of the life cycle as one of the dimensions in a Modular Engineering. Life cycle consideration gives the demand to see modular systems being dynamic according to certain rules. Above mentioned research programmes: *Intelligent Manufacturing Systems* (IMS) and *Global Engineering Networking* (GEN) give possibilities to make modular systems more dynamic. GEN supports engineers and companies by being also an electronic marketplace for modules and as the medium for virtual enterprise.

Our definition for Dynamic modularisation is:

Dynamic modularisation is the novel Modular Engineering process, which allows bringing in a dynamic way new more merited modules to the system, and leaving out the old ones. This process is based on the definition of the encapsulation, similarities and the description of interfaces as well as modular management system. All different stakeholders' views should be taken into account; other dimensions will be very similar to those defined for modularisation.

Dynamic modularisation can change the design process. It is a new way to classify and utilise the organ structure. According to the Theory of Technical

Systems by Hubka & Eder [14] and Domain Theory by Andreasen [13] there exists some classes of functions and organs, in accordance with the so-called *wirk-complex law:* "The realization of a working effect needs realization of energy, helping, control and supporting effects."

We may expect that a modular system consist of module classes that are:
- Working Modules
- Auxiliary Modules (energy, helping, control and supporting modules)
- Secondary Modules: Connecting Modules

In some classes of machines, especially when modules are of the transformation type, there exist secondary modules, which may be called connecting modules.

8.2
Simplified Modular Structure

A machine or a technical product consists of three types of units: working units and their subunits; connecting and frame units. Same differentiation to three can be made in modular system: to working, connecting and frame modules.

When the frame modules solve supporting tasks (i.e. transmission or inhibition of force, heat, vibration or noise), connecting modules solve the transfer of operand (material, energy and signals). The working modules are carrying out a diversity of tasks: working, energy, helping and control.

Only transformation defined modules may be treated in this way. Transformation module may be defined by its relations, structure and behaviour. We may also add organisational attributes like origin, supply etc. Based on information of attributes it is possible to search suitable, optional modules for a product, which is under development.

8.3
Design Process in Dynamic Modularisation (DYMO)

In Hubka's theory a design is controlled by two main dimensions:
1) Vertical causality chain - Organ structure
2) Horizontal causality chain - Transformation process.

Sometimes the vertical causality chain consists only of functional relations like forces and moments, sometimes it consist of a transformation chain, for instance of changing energy or signals for control.

The design activity should start with the horizontal causality chain and then build up the vertical chain, which are auxiliary functions and their transformational or functional relations and their in/outputs of energy, materials (e.g. water or oil) and signals (digital/analog).

In transformation to DYMO we will change the design process to start from the most important area: the realization of the process. The last design task will be the design of the frame.
1) Design starts from the specification. The transformation process schema will be developed.

2) In the future the process schema gives a draft Virtual Reality Modeling Language (VRML) -models of modules and their connections. For the space model of a product there are some constraints. Based on this starting situation the first 3D model of modules will be structured. As VRML models modules are easy to transfer and rotate. The different spatial modules are fast to study.

3) The next step is to add connecting modules to the model. Some critical connectors should be minimized and all connectors must be reasonable in length. Getting information of the selected connectors as a table form and comparison to constraints could make this optimization.

4) Finally when the space model of modules and connections is designed the design of a frame can be started. The frame causes some iteration for the placement of the modules and connectors.

We will test this design process in mobile machinery industry and in power plant projecting business.

9 Evaluation and Configurability

9.1
Evaluation in Configuration

Evaluation is a necessity for choosing the optimal solutions from optional variants. It also is needed in the Product Family Development stage for analysing whether developed propositions are suitable or not for profitable configuration.

When the system product will be configured within Configuration Management system, an evaluation system is needed to compare different options. In Department of Control and Engineering Design at Technical University of Denmark researchers have applied the seven universal virtues: Cost, Quality, Flexibility, Risk, Lead time, Efficiency and Environmental effects for comparing different options. Engineering Design Centre at Cambridge University has developed an approach to the configuration optimization and a term of a technical merit [19]. The technical merit of a system is determined by its position in the design space relative to comparable systems, which have been found to lie on the constraint boundary. A preliminary method can be developed for calculating the indices of duty (performance), reliability and cost for existing solutions. The technological constraints of possible alternative solutions, however, must be forecasted. The chart for the technical merit is presented in Fig. 10.

9.2
Configurability

In October 1997, a project for creating a methodology and computer support for modeling configuration knowledge was started by industry and two research institutes: Institute of Machine Design at Tampere University of Technology and Product Data Management Group of TAI research institute at Helsinki University

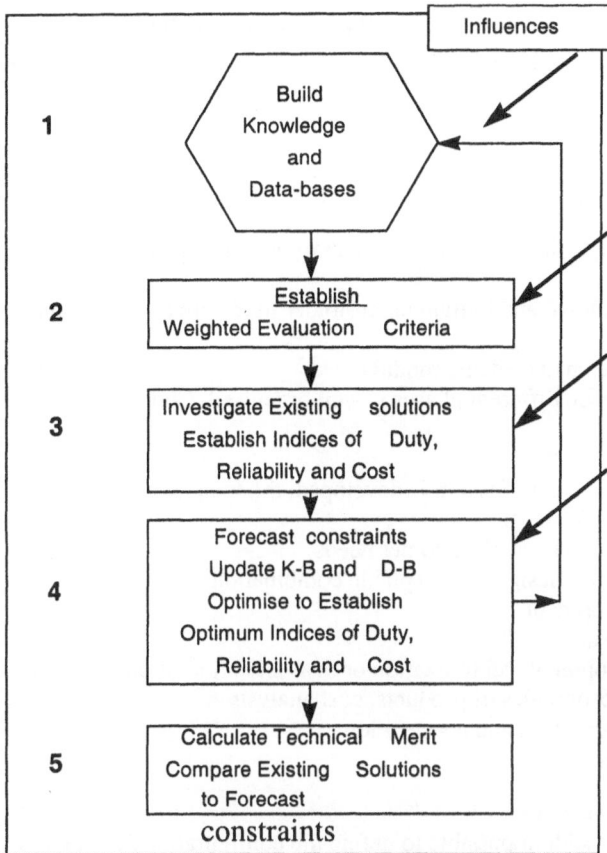

Fig. 10. Flow chart for the calculation of technical merit [19]

of Technology. The research on methodology included literature research, interviews in companies, and systematisation of knowledge into a form of configuration principles. The computer support part of the research consisted of creating generic configuration knowledge modeling (computer) language and a computerised method for analysing configurability of a version of a product under development.

Several concepts were developed in the project for an evaluation of configurability. The most important criteria are following [15].

1) The product programme has to show *variety*, which is interesting for the market, i.e. such types of differences concerning functionality, yield, qualities and features that enable one to create different types of businesses and big turn-overs.

2) The product programme must show the lowest possible *complexity* in all activities, which relates to product development, manufacture and life cycle activities.

3) The product programme must show the highest possible degree of *commonality* in relation to the systems found in the manufacturing and product life cycle.

10 Configuration Management Principles

According to our experience, the successful configuration demands the following three prerequisites.
1) Customer needs are known and individual customer needs are classified and formed into groups
2) Products are analyzed, managed and modularised
3) Communication between different phases is well organised, Concurrent Engineering

When Configuration Management System is developed, the focus should cover one or many of the following areas:
- Sales support system, analysis of customer needs
- Product analysis, the pre-design of acceptable combinations
- Documentation development
- Workflow development
- Subcontracting development; centralization or diversification of outsourcing
- Value management, economics of products, cost analysis
- Configuration of organisation and cooperation
- Feedback compilation

Principles for the development are:
1) In the future it is only seldom possible to define the configuration system as a closed system, a new technology has to be transferred to the system
2) The configuration of the product has to be made covering the whole life cycle.
3) The Configuration Management System has to include all the structures of the Domain theory
3a) the organ structure of the product offers a possibility to manage product in long term
3b) the function structure is important for sales.

11 Conclusions

In this chapter we have presented many different views to Configuration Management. This presentation is based on the R&D projects, from which Tampere University of Technology has gathered practical experience for years. In our projects with industry we have got good results. However, we have found that we have to return back to some research areas, which we thought we were managing. More research is still needed. We refer to research areas like Knowledge Management, alternation between Process, Organ and Assembly-structures, and Dynamic Modularisation. Inside these headlines there are many interesting research questions. We will continue research work and industrial cooperation on these areas. The 5th WDK workshop gave us a plenty of new sights, new ideas and inspiration to continue the configuration research.

12 References

[1] Pine Joseph: Mass Customisation: The New Frontier In Business Competition.
[2] Brian Maskell Software And the Agile Manufacturer: Computer Systems And World Class Manufacturing.
[3] http://www.c-lab.de/gen/
[4] http://www.ims.org/
[5] Mittal, S., Frayman, F.: Towards a Generic Model on Configuration Tasks, Proceedings of the Eleventh IJCAI, pp1395 – 1401
[6] http://www-cdr.stanford.edu/SHARE/share.html
[7] Paasiala, P.: An Approach to Design of One-of-a-Kind Products, Draft for PhD thesis, 1997, Tampere University Of Technology
[8] http://www.trilogy.com/Pages/Solutions/Ebusiness/Configuration.htm
[9] http://www.i2.com./
[10] TEKES: Improving Product Development Efficiency in Manufacturing Industries 1996-1999, Technology Programme Report 3/2000, Final Report, ISBN 952-9621-63-9, 2000, Helsinki, 38-142
[11] Tomiyama, T.: Concurrent Engineering: A Succesful Example for Engineering Design Research, 1st International Engineering Design Debate (EDD '96) Gasgow, september 23-24,1996
Mørup, M.: Design for Quality, Diss. Institute for Engineering Design, Technical University of Denmark, 1993.
[12] http://www.km.org./gwa/papers/NASA.html
[13] Andreasen, M. Myrup: Methods for Synthesis, a System's Approach (in Danish). Diss. Lunds University of Technology 1980.
[14] Hubka, V., Eder, W.E.: Theory of Technical Systems, Berlin/Hedelberg& New York, Springer-Verlag, 1988.
[15] Riitahuhta, A. & Andreasen, M. Myrup: A theory of Structuring of Product Families. Department of Control and Engineering Design, DTU. (Planned).
[16] Andreasen, M. M., Hein, L.: Integrated Product Development,
[17] Olesen, J., Wenzel, H., Hein, L., Andreasen, M. Myrup: Design for Environment (in Danish), Miljø- og Energiministeriet 1996, ISBN 87-7810-435-1.
[18] Laurikkala, H., Tanskanen, K., Nevalainen, P., Vainio-Mattila, M.: Consortium as a Virtual Enterprise in Project Planning, Proc. 5th Int. Conf. on Concurrent Enterprising, March 15-17, 1999, The Hague, The Netherlands.
[19] Murdoch, T.N.S., Wallace, K.M.: An Approach to Configuration Optimization, Journal of Engineering Design, Volume 3, Number 2, 1992, pp. 99 – 116

Workshop Structure and Comments from Discussion

Antti Pulkkinen

Tampere University of Technology
P.O.Box 589
33101 Tampere, Finland
e-mail: pulkkine@ruuvi.me.tut.fi

Abstract. In this section, the workshop framework, working procedure and the contents are examined. Short overview on the papers and presentations is given. The framework is based on procedural view on Design for Configuration. Two other structures for the contents are presented. These are supposed to help the reader to link the papers to product structuring research and to each other.

1 Introduction

The Laboratory for Production Engineering, Delft Technical University, has arranged the previous WDK Workshops on Product Structuring in Delft, the Netherlands. This time the workshop took place in Tampere, Finland, and the arrangements were carried out mainly by the Institute of Machine Design, Tampere University of Technology.

In earlier workshops there had been:

- a number of different presentations (each followed by discussion on the issues related to presentation)
- a concluding remarks for both of the days
- conclusion for the workshop itself

Apart from giving the presentations, the rest of the activities had been conducted by a small group of enthusiastic professors. In the group, there usually had been a person to make the concluding remarks.

It was decided to increase the activity of the participants. This was suggested to happen by having more time on the discussion and introducing the structured brainstorming method on the discussions. However, a lot of the good practices from the earlier workshops were not discarded. Among these practices were giving a special theme for the workshop (the practice was applied also in 1998 workshop) and having short discussions after each presentation. We also made concluding remarks after the workshop, and provided the participants a structured framework beforehand so that they could relate the presentations to overall framework.

2 Workshop Structure

In call for papers the theme for the workshop was stated as "Design for Configuration" and the framework for the theme was given.

We expected that the framework would be broad enough to cover all the contributions. The topics covered by the framework were recognized as important ones in the research project we carried out during years 1998 and 1999 with four Finnish companies and TAI research institute from Helsinki University of Technology [1].

After the papers had been received and the evaluation had been carried out, it was recognized that no one of the authors had specifically addressed the management of dynamics of product portfolios or platforms, module systems, and modules. Thus, the topic area was discarded from the topic list and the workshop was concentrated to four sections (see Table 1).

Table 1. Sections of the workshop

Analysis of customers, markets, and technology	**Metrics and methods** for modularity and configurability
Development of product portfolios and module systems	Supporting **modeling and IT-tools**

2.1
Double Team Group Work

In the call for papers it had been emphasized that "to get the most out of the gathering we plan to arrange as much group work as possible". To make this possible, we had drafted a version of double team brainstorming method for the workshop. A smilar method has been earlierly used in Concurrent Engineering research by Lettice et al. [2]. The method consisted of four sections (see Fig. 1).

The *individual work* was supposed to take about five minutes. In it, all the 32 participants were asked to register ideas (put notes on the paper) about the topics of the section. They were guided to *concentrate on essential aspects* on the specific subject – e.g. the *goals and problems* of analysing markets, customers, and technologies. The participants were also asked to think about the *methods* to achieve the goals and to avoid the problems of the topic and make observations on the *criteria* on succeeding in the topic area – e.g. what are the goals, problems, and criteria in developing modular product portfolios for configuration

Then the participants were supposed *work in pairs* to collect and compare the results of individual work. From this phase of group work we expected to get more comprehensive and refined opinions on the area. In the schedule there was ten minutes reserved for making the comparisons in pairs.

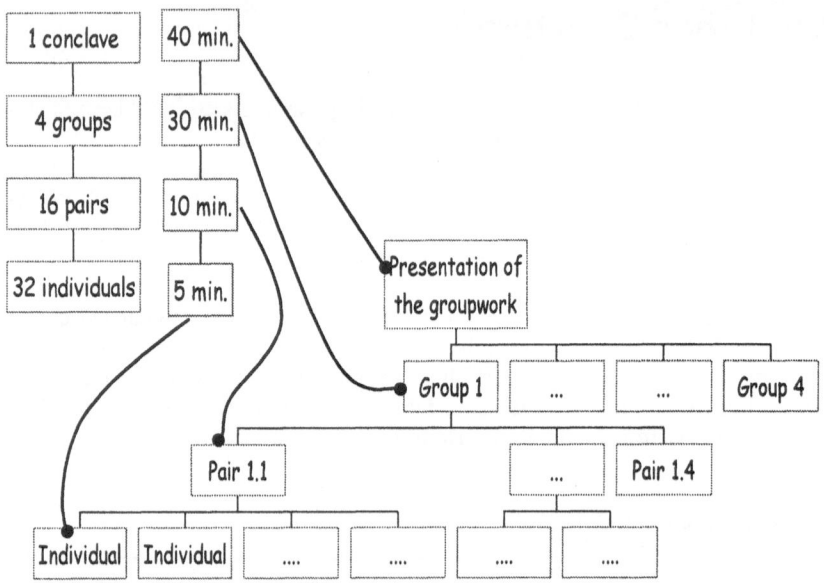

Fig. 1. The structure of the groupwork

About 30 minutes was scheduled for the actual *group work*, where the pairs told about their notions on the subjects. These notions were combined and collected to presentations, which were given right after the group work. About 40 minutes was to be used for the four *presentations of the group work*.

Group work took place right after the presentations of each section. In the workshop schedule there was six hours reserved for brainstorming sessions. Thus, by having the four group work sessions, we expected to create almost three hours of parallel discussion in groups (see Fig. 1). Since the schedule was rather tight, the last session had to be cancelled. A short discussion about the contributions and conclusion about the future research topics in the field replaced it.

3 Workshop Contents

Both days of the workshop followed the structure presented in Fig. 2. In the first day there was a formal opening by professors Asko Riitahuhta and Jarl-Thure Eriksson (the headmaster of Tampere University of Technology). In the second day the closing session (led by Marcel Tichem from Delft Technical University) replaced the brainstorming session.

The actual topics of the workshop was opened by Niels Henrik Mortensen, who gave an introduction under a titled *The need and proper understanding of modularisation*. It included notes on importance and relevance of modularisation, how modular engineering is related to business, which person (in a company) should be responsible for making decisions on modularity, and some open questions.

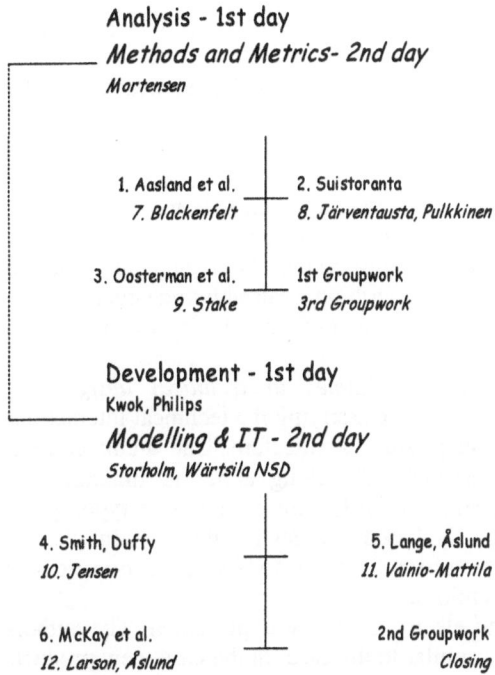

Analysis - 1st day
Methods and Metrics- 2nd day
Mortensen

1. Aasland et al.   2. Suistoranta
   7. Blackenfelt   8. Järventausta, Pulkkinen

3. Oosterman et al.   1st Groupwork
   9. Stake   3rd Groupwork

Development - 1st day
Kwok, Philips
Modelling & IT - 2nd day
Storholm, Wärtsila NSD

4. Smith, Duffy   5. Lange, Åslund
10. Jensen   11. Vainio-Mattila

6. McKay et al.   2nd Groupwork
12. Larson, Åslund   Closing

Fig. 2. Sections, presentations and groupwork

3.1
Analysis

In a paper *Even Modular Products Can Be Unmanageable* by *Aasland et al* [3] there is an initial description of a business situation for a research project. In the situation a company has done all the right things, i.e. done everything "by the book", in making a modular product program. Albeit this, the company has ended up with an unmanageable amount of sales items and a sales method that does not seem to be the best possible way of practice. To solve the problem, writers suggest a concept of an "extended product", i.e. a mindset where both the product and the services related to it are taken into consideration as a market, sales, product development, and production item. They also recommend that customer behaviour in sales situation and understanding customers' psychological preferences in these situations should be studied and taken into practice in designing product platforms and modular product programmes.

During the discussion, that came right after the paper, it was stated "the company might not have had the right understanding of the customer preferences in the beginning" and "they may have not understood the difference between marketing and sales situations". This was extended as it is very important to have a proper understanding of the customer needs, their commonalities and variations, when aiming at modular product programme and offering a variety of product configurations to markets.

Suistoranta [4] takes an effort to combine the complicated needs of sales situation in a globally operating projecting company and academic means to describe the situation. The academic means are taken from marketing theory and theory of technical system. The theories are applied for describing the business domain definition, concept selection development, and configurator development. These phases form a concept of *Structured Sales Processes for Configuration*.

After the presentation there was a short discussion about the importance of having the elements and their relations for describing a sales/buying situation.

Nowadays it seems that there is no general way to describe this essential situation of configuration.

When the paper *Visualizing interdependencies in technical design* was presented, the focus turned on describing and classifying the technical interactions by having an analysis of relations in product architecture. The architecture is spelled by having constraints[1] on entities that belong either to functional or physical domains. In the paper *Oosterman et al* [5] are presenting a typology of technical interactions. The typology is based on having three main classes: interactions between functions, mapping form functions to physical entities (called chunks), and integration of physical entities.

In the presentation a case with an electric shaver was presented. The authors suggest benefits for design processes similar to the case. In the case, new products are developed on the basis of the existing product architecture and designers are expected to improve product through manipulating the integration of physical entities. The structuring of product development organization should support the interactions.

3.2
Development

The second section of the first workshop day consisted of industry opinion on *Development of product portfolios and module systems* and related presentations. Also a brainstorming section on the topic was included.

Sonny *Kwok*, from Philips Garment Care Singapore Pte Ltd, which is a leading company in the field, gave the introduction Changing Paradigm. Kwok stressed that the need in the company is to reduce time to market of customized products. In the field the product renewal rate is high while the product innovation rate is low. Hence, company's product portfolio represents mature technology in turbulent markets.

In the presentation, it was suggested that product architecting is a means for compressing time to market. The company's business objective is flexibility in an easy, cheap and fast way. This is to be enabled through creating a breakthrough

---

[1] Actually constraints are combinations of energy, material, and information relations between entities and spatial constraints in physical domain.

product architecture that optimizes between industrial commonality and commercial diversity. The project for reaching the architecture consists of 6 sub-projects as follows:

1. Identify existing situation and constraints.
2. Draw Architecture based on wishes from stakeholders.
3. Build Awareness and Clear deployment in organization.
4. Make clear endorsement on decision: how to proceed
5. Create & Apply framework/process/procedure
6. Review & Evaluate actions

Smith and Duffy [6] had prepared a presentation on the topic *Product Structuring for Design Re-Use*. The authors provide a literature study on general benefits of re-using design knowledge with a formal way, a classification of the types of re-usable knowledge, and a description of what is the usual re-use way of work and a general description on how re-use should be formalized. Another part of the paper presents knowledge structuring in relation to re-use requirements, suggest an organizational structure for re-use, and presents a design knowledge re-use process model mapping to product structuring research.

The authors conclude that some areas of the mapping are not supported by current research. Their main concern seems to be the small amount of research on domain exploration and design for re-use. Writers point out also the need of methods to support acquisition of abstract and contingent design knowledge.

In the discussion, it was noted that some informal methods for enhancing re-use should be researched. It seems that usually the structures of knowledge are leading to ever more complex data- and knowledgebases. In practice there should be guidelines and/or methods for deciding about the data and knowledge that is to be re-used and computerized.

Presentation on *Capturing Quality Perceptions in the Design Rational of a Modular Product Concept* was given by Mark *Lange* [7] (Johan *Åslund* is the co-author). The paper presents a mindset of different quality levels and classification on quality attributes. These are based on semiotic sign theory. The result is an approach where designs are apprehended as sign vehicles having a meaning in observer's mind. The typology leads to different levels of content in a semiotic design space.

In a case study, these attributes are interlinked to a design synthesis description. The description is based on MFD (Modular Function Deployment), which Erixon has presented. Finally the authors give an evaluation on different kind of quality attributes in a modular product and make some remarks on the distribution of quality properties in product architecture.

In *Relating Product and Process Structures, McKay et al.* [8] suggest an approach where not only product data is being structured but also the process data is structured by using similar concepts. The authors give a general data architecture for expressing both product and process related data entities and their relations. This is based on the experiences from developing STEP standard (ISO 10303) Data Specification Architecture.

In the presentation it was emphasised that the suggested data architecture is supposed to support only the way for modelling the different data items in a uniform way. It is supposed not to give hints or suggestions of what should be modelled and certainly not to give methods to support the structuring of products and related processes – the architecture is for only the related data.

3.3
Metrics and methods

The second day of the workshop begun with third section *Metrics and Methods for modularity and configurability*. The section included an introductory case study and three presentations, which were followed again by a brainstorming session.

Mortensen's case study presentation was about *a Procedure for modeling product families in configuration systems*. In the introduction, configuration and its meaning to a company were described. Mortensen gave a note of the plausibility of configuration as a business solution to certain kinds of companies. He described the impact of having a configurator by effects of solving tasks with higher frequency and complexity, linking structure and behaviour of a technical system, focusing on critical and competing edge elements, and reducing the resource consumption. These effects lead to a rationalization of product sales-delivery process, where the benefits of configurator are harvested.

Mortensen described configuration system, which is based on modules, assemblies, parts, etc., as an entity containing the knowledge needed in configuration. In practice, the development of the system is a substantial part of product development process. However, the process is lacking conceptual phase and existing modeling methods supported by configurators do not support conceptualization very well.

For conceptualization Mortensen suggested a procedure consisting of five consecutive phases. These were definition of configuration task, identification of product family master plan (PFMP), conceptual modeling of PFMF, detailed modeling of PFMP, and coding & testing phase. Mortensen stated that the procedure, the PFMP and consideration of the life-cycle aspects enhance the spelling of product assortment into configuration model, which is to be used in configuring. The results of a case study, which used the procedure, were promising.

A paper *Modularisation by relational matrices – a method for the considera-tion of strategic and functional aspects* was presented by Michael *Blackenfelt* [9]. In the paper, two known methods, Module Indication Matrix (MIM) and Design Structure Matrix (DSM), are treated. The former is used in presenting the strategic and the latter the functional aspects of modularity. Furthermore the strategic aspects, indicated by module drivers, are merged with the DSM method. To enable this a clustering of module drivers is suggested. Also some heuristics and metrics for clustering matrices are presented in the paper.

Blackenfelt uses an earlier case study (on vacuum cleaner) to demonstrate the effect of merging the different aspects into a more comprehensive method. The heuristics are used in re-organizing (partitioning/clustering) both the functional and the strategic DSM's of the case. Analyzing published case tested the metrics. Suggestions for rival product architecture in the case were found (i.e. new module structures are identified).

The section's second paper *Enhancing product modularisation with multiple views of decomposition and clustering*, by *Järventausta and Pulkkinen* [10], is also about using matrix methods in product structuring. The authors present a method for clustering product data items according to their architectural relations. Different stages of clustering are supposed to form the hierarchical (part-of/is-a) relations between items. Product data management (PDM) system is suggested as an IT-tool for viewing the clusters. The proposed points of views are marketing, design, and production views.

A case study for testing the approach is presented. The topic is the functionality of a mobile rock-drilling rig. In the case, a matrix of the product data items is generated from a chart. In the chart, functional units of a high-resolution level are related to each other. Then the units and their relation are transformed to a symmetric DSM. A procedure for clustering is presented and a well-known algorithm is used in the procedure. As a result the functional composition is produced according to the interrelations between sub-functions. Writers give some remarks on benefits and applicability of the presented method. They also note the ability to use automation (i.e. clustering algorithms) in product structuring.

In *A framework for evaluating commonality*, Roger *Stake* [11] gives a literature review on the product families and platforms, commonality, and measuring methods for commonality in product family. He specifies two characteristics of measuring parameters, i.e. 1. pragmatic and 2. analytical parameters. These are further classified to be used from two points of view: analyzing the structural level (i.e. commonality of modules and components) and analyzing the architectural level (i.e. product family and variant commonalities).

Stake points out that by using different methods the results of commonality indices the measurer gets very different values. Consequently, it is important to have a good understanding on the measurement rationale, when a certain method is applied. Also, there seems to be different requirements for the measured data and, hence, different methods are suitable for different phases of product architecture development.

3.4
Modeling and IT-tools

Jensen [12] gave a presentation on *Preservation of Engineering Knowledge in Configuration Systems*. It was based on the fact that engineering knowledge is dynamically evolving, while the product configuration model tends to be static one. Jensen presented a typical example for illustrating this consistency problem. Furthermore, he suggested that the abilities of current configuration systems are limited for modeling artefact on one (structural) domain only. This is usually not enough for expressing the needed views of a configuration. Jensen suggests that it is especially important to have the ability to provide selections on functional domain for customers.

In the paper Jensen suggests a module interaction object to be used in configuration knowledge modeling. This object is supposed to ease the maintenance of configuration model (i.e. the uploading of a new module to the configuration system). Also different domains and mapping between them should be provided.

A case on milling machine interaction object (a cone joint) is presented. In it the attributes of cone joint are related to formula covering the geometric and functional (torque transmission) constraints. The reduced effort of adding new modules for configuration system is drafted by comparing the systems with and without interaction object. Finally, Jensen gives remarks on the location of product configuration knowledge among software systems in a company.

Report on Using Product Data Management Software in Managing Modular Products was presented by *Vainio-Mattila* [13]. The report is based on observations among five companies that are having modular product architecture as a goal and a PDM system as a means. The report is an initial survey for discovering the needs, that modular product architecture sets for PDM systems, and the possibilities that current PDM systems offer for management of modularity.

After the giving the results of the survey, Vainio-Mattila emphasizes two issues. Firstly, the creation of specifying data and data management process is needed in a PDM implementation project. Secondly, the trade off of modeling has to be clarified and related decisions have to be made. Remarks on benefits / efforts are made on the end of the paper.

A tool for representing relation between costs and product architecture was presented by Tobias *Larsson* [14] (Johan *Åslund* was a co-author of the paper). The paper ValueMap™ – a method for understanding the economical potential of product modularization and cost of variety distuinguishes different stakeholder objectives in different organisational levels and life cycle span. The presented tool is mainly targeted for management to represent the benefits of having a modular product architecture.

The tool is based on Activity Based Costing (ABC). In it, the cost objects (e.g. indirect labour, machines, real estate) are related to activities of a company. The activities are then related to cost drivers (e.g. part number, variant, base model, volume). The results are percentages of Product Structure related, Production Volume related and Micellaneous costs. The procedure supported by the tool is presented by using a case in a Swedish company.

The authors give a generalisation on the costs relations to product structuring. They also indicate the limitations of current application of the tool and relate the tool to concept evaluation and management support processes.

4 Another Views on the Contents

From structural or architectural point of view the papers are individual objects, but as in science in general the papers have to be related to existing research. This research can be stated in two dimensions. First, the overall issues on the specific field, which is product structuring. Second the relation to each other. The second relation is because the papers are contributing to the science itself. In the following these aspects are studied.

4.1
Papers in Relation to Key Issues of Product Structuring

The papers covered a large variety of research fields. However, as Mortensen stated in the opening presentation the three key elements of product structuring are:

1. Product / hardware[2] itself (configurations, modules, components, parts, etc.)
2. Knowledge, Information, and Data
3. Activities, e.g. tasks, organizational processes, and life-cycle phases

In Fig. 3 the papers are arranged according to their relation to these three elements of structuring. The arrangement is based on my insight, which took shape during the evaluation of the papers.

According to the figure the first group would be the papers that are mainly targeted on relating the first two issues (the product and information items). The papers by Smith & Duffy, Järventausta & Pulkkinen, Vainio-Mattila and Jensen [6, 10, 12, 13] are clustered into this group.

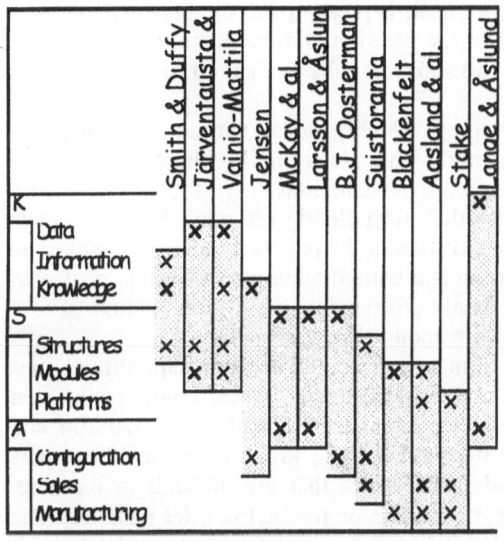

Fig. 3. The papers according to the key elements related to product structuring

The second group is composed of the papers that relate the pure activity and product structure issues. Among this group I counted in the papers by McKay et

---

[2] It may be arguable to use the word *hardware* here, because the components of the product might as well be composed of software components. Product technology, which sets the constraints for activities and knowledge, is referred with hardware.

al., Larsson & Åslund and Oosterman et al. [5, 8, 14]. These papers deal the relation in higher and more abstract level than the papers of third group.

The third group is relating some of the elements of first issue (structures, modules, and platforms) into some of the elements of third issue (configuration, sales, manufacturing). This group is composed of the papers by Suistoranta, Blackenfelt, Aasland et al. and Stake [3, 4, 9, 11]. The paper by Lange and Åslund [7] was touching mostly the elements of second and third issue. So, it belongs to a group of its own.

4.2
The Papers in Relation to Each Other

A second insight during the evaluation period was developed on the papers in relation to each other. The insight was based on the used methods (e.g. DSM was mentioned in the papers by Blackenfelt, Järventausta & Pulkkinen, and Oosterman et al.) and covered topics (e.g. the role of the papers presented in Fig. 2 and paragraph 4.1). Thus, the insight is supposed to collect both the views on DFC framework and related research.

To demonstrate the second insight, a DSM (see Table 2) on the papers and their relations was derived. Because the tool itself is presented in the papers previously mentioned, the detailed explanation of the creation and the meaning of DSM are left out from here. It is based on remarks made on the evaluation form, which was filled for each paper.

The matrix above presents a grouping, which is made without any clustering algorithm (i.e. by trial and error method). This clustering highlights four groups that are somewhat different from the earlier presentations.

In the first group, there are papers that most clearly are related to knowledge structuring issues. The group is a composition of the papers by Smith & Duffy and Jensen. The second, rather large, group is a combination of papers that deal with modularisation, DSM and product data management. It's a combination of five papers (Oostermans paper is shared with another group).

The third group is a combination of management and decision supporting issues in product structuring: interaction and organizational issues, commonality and platform metrics, and cost analysis. Oosterman et al., Stake, and Larsson and Åslund shape the group from three papers. Last (4th) group combines the papers that are related to management and 'soft' issues that are difficult to measure. These are mostly related to customer perception on product in sales situation. The group is a combination of papers by Suistoranta, Aasland et al., and Lange & Åslund [3, 4, 7].

5 Discussion on Structuring the Workshop Contents and Conclusion

As presented the presentations and papers can be observed from many different points of view. In this introduction, three ways to look the contents of the workshop are given. These are:

Table 2. Clustered DSM on the papers

	Smith & Duffy	Jensen	Blackenfelt	Järventausta & Pulkkinen	McKay & al.	Vainio-Mattila	Oosterman et al.	Stake	Larsson & Åslund	Suistoranta	Aasland & al.	Lange & Åslund
Smith& Duffy	x	1										
Jensen	2	x										
Blackenfelt			x	2			1					
Järventausta& Pulkkinen			2	x		2						
McKay & al.					x	1	1		1			
Vainio-Mattila				2	1	x						
Oosterman & al.			1	1	1		x		2			
Stake	1		1					x	1			
Larsson & Åslund								1	x			
Suistoranta	1			1			1			x	2	1
Aasland & al.									1	1	x	
Lange & Åslund										1	2	x

- Procedural viewpoint
- Product structuring research
- Arrangement of papers

The mindset behind the chosen workshop structure (procedural viewpoint) was based on observations on development of a product family for configuration. Generally, the project is twofold (see Fig. 4). First, these processes begin with a product structuring/platform development project. Second they continue, by module development projects. These projects can be done partly in parallel, but a certain degree of platform development is necessary for module development.

Andreasen and Hein have developed a model for Indegrated Product Development. In the model there are six phases where members of development team from market/sales, design, and production deparments have specified activities. Similarly to the IPD model, the framework of Design for Configuration begins with analysis. As in IPD model there are three main stakeholders, also in the framework the projects begins with analysis from both the sales, product and production point of view.

The IPD model continues with the actual development procedures. The framework continues similarly, but contains sections for metrics and methods, and modeling and IT support.These are important for supporting the development process and evaluating and using the results. Thus, the framework should be at its best providing the viewpoint to practitioners.

The second structure for the papers is a more theoretical one (see section 4.1). This approach provides information for researchers from structuring vs. knowledge or structuring vs. activities the points of view. Similarly to second structure the third one is giving a more research-oriented point of view to the papers. However, the third structure shows how papers are related to each other. Thus, it is a combination of viewpoints. For instance, it merges theoretical approach of knowledge structuring to the practical possibilities of solving the problems of knowledge modeling in current configurators.

Since the purpose of the workshop proceedings is supposed to support the Design for Configuration process, the framework structure was selected for the proceedings.

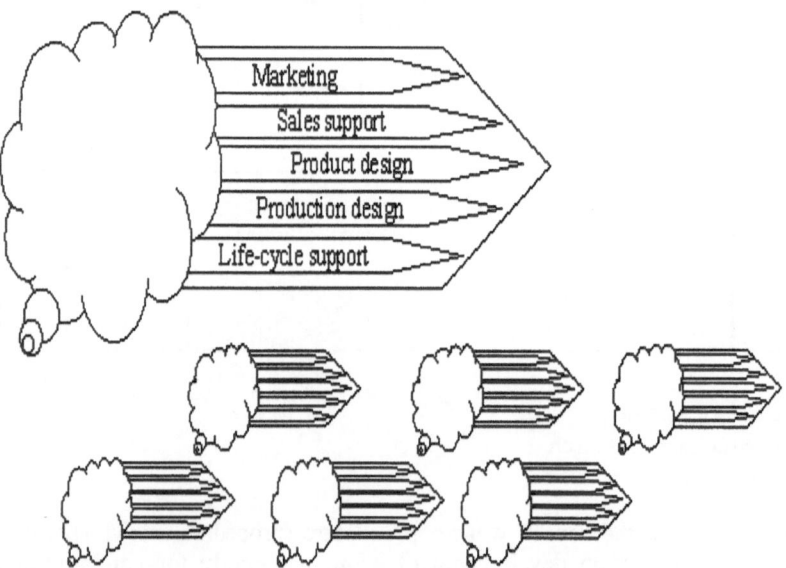

Fig. 4. The structuring project vs. module development projects (from [1])

6 References

[1] Pulkkinen, A., Lehtonen T., Riitahuhta, A. Design for Configuration – Methodology for Product Family Development. In: *Proceedings of ICED 99*. Edt. Lindemann, U. et al. Technische Universität München 1999.

[2] Lettice, F., Evans, S., Smart, P. Understanding the Concurrent Engineering Implementation Process – A Study Using Focus Groups. In: *The Design Productivity Debate*. Edt. Duffy, A.H.B. Springer-Verlag. 1998.

[3] Aasland K., Reitan J.,Blankenburg D., Even Modular Products Can Be Unmanageable. *Proceedings of the 5th WDK Workshop of Product Structuring 2000*. Edt. Riitahuhta A., Pulkkinen A., Springer-Verlag 2001

[4] Suistoranta S., Structured Sales Process for Configuration, *Proceedings of the 5th WDK Workshop of Product Structruring 2000*. Edt. Riitahuhta A., Pulkkinen A., Springer-Verlag 2001

[5] Oosterman B J., Gaalman G J C., Kuijpers F P J., Visulizing Interdependecies in Technical Design, *Proceedings of the 5th WDK Workshop of Product Structruring 2000*. Edt. Riitahuhta A., Pulkkinen A., Springer-Verlag 2001

[6] Smith J S and Duffy A H B, Product Structuring for Design Re-use, *Proceedings of the 5th WDK Workshop of Product Structruring 2000*. Edt. Riitahuhta A., Pulkkinen A., Springer-Verlag 2001

[7] Lange M., Åslund J., Capturing Quality Perceptions in the Design Rational of a Modular Product Concept, *Proceedings of the 5th WDK Workshop of Product Structruring 2000*. Edt. Riitahuhta A., Pulkkinen A., Springer-Verlag 2001

[8] McKay A., de Pennington A., Trott S J., Relating Product and Process Structures, *Proceedings of the 5th WDK Workshop of Product Structruring 2000*.Edt. Riitahuhta A., Pulkkinen A., Springer-Verlag 2001

[9] Blackenfelt M., Modularisation by Relational Matrices – a Method for the Consideration of Strategic and Functional Aspects, *Proceedings of the 5th WDK Workshop of Product Structruring 2000*. Edt. Riitahuhta A., Pulkkinen A., Springer-Verlag 2001

[10] Järventausta S. and Pulkkinen A., Enchancing Product Modularisation with Multiple Views of Decomposition and Clustering, *Proceedings of the 5th WDK Workshop of Product Structruring 2000*. Edt. Riitahuhta A., Pulkkinen A., Springer-Verlag 2001

[11] Stake R., A Framework for Evaluating Commonality, *Proceedings of the 5th WDK Workshop of Product Structruring 2000*. Edt. Riitahuhta A., Pulkkinen A., Springer-Verlag 2001.

[12] Jensen T., Preservation of Engineering Knowledge on Configuration Systems, *Proceedings of the 5th WDK Workshop of Product Structruring 2000*. Edt. Riitahuhta A., Pulkkinen A., Springer-Verlag 2001.

[13] Vainio-Mattila M., Report on Using Product Data Management Software in Managing Modular Products, *Proceedings of the 5th WDK Workshop of Product Structruring 2000*. Edt. Riitahuhta A., Pulkkinen A., Springer-Verlag 2001.

[14] Larsson T. and Åslund J., ValueMap™ – a Method for Understanding the Economical Potential of Product Modularization and Cost of Variety, *Proceedings of the 5th WDK Workshop of Product Structruring 2000*. Edt. Riitahuhta A., Pulkkinen A., Springer-Verlag 2001.

Part I

Analysis of Customers, Markets and Technology

Record from 1st Group Work

Antti Pulkkinen

Tampere University of Technology
P.O.Box 589
33101 Tampere, Finland
e-mail: pulkkine@ruuvi.me.tut.fi

1 Introduction

First the participants were introduced the method. They were advised to concentrate on analysis of customers, market and products. Results from group work were presented and notes on them were made. Some common topics of discussion were found.

Most of the participants emphasised that the new way of work requires new procedures, management style, thinking patterns, etc. These are discussed in following section. Also a question on what are the needs for different methods/ means was addressed.

2 Records from Discussion on Analysis

It is vital to have a framework on what are the elements of analysis for configuring and how they are related. What is a module and what is a good one? How to structure activities and organisations? How to isolate product architectures from organisational changes? Does product concept drive platform development? From where to start?

Those who deal with product structuring need procedures and methods for determining required subjects of variation. In practice this means the rational way for "where to put the variation" and requires decision support for which alternatives to offer. This calls for better understanding of markets, but are there any suitable method for that? Portfolio Management and Quality Function Deployment (QFD) have been earlier suggested as preferable methods, but are they really applicable with product platform development, modular engineering, and configurable products. What would be the overall marketing approach? What is a general goal and does there exist just a one goal?

2.1
Knowledge Capturing and Re-use

Product development deals with knowledge of needs and possibilities to satisfy the needs. It was stated in most of the group work presentations that there is a challenge in understanding both the expressed and unexpressed needs (i.e. customer requirements). Capturing the knowledge on unexpressed needs is a demanding task.

More generally, the problem is related to types of knowledge. When explicit information and context tends to be related to explicit knowledge, implicit information or context is related to tacit knowledge. Handling the knowledge is becoming more important and challenging issue. Product knowledge modelling enhances usually capturing knowledge for design re-use, but how to model tacit knowledge?

2.2
Challenge in Understanding Customer's and Managing Products

The problem of analysis is a composition of several issues. It includes both the questions "how to make the analysis" and "what to analyse". In the workshop the emphasis was on analysis of customers, because "customer needs form the basis in analysis".

What really are the customer perceptions and how does she/he experience the critical sales situation?

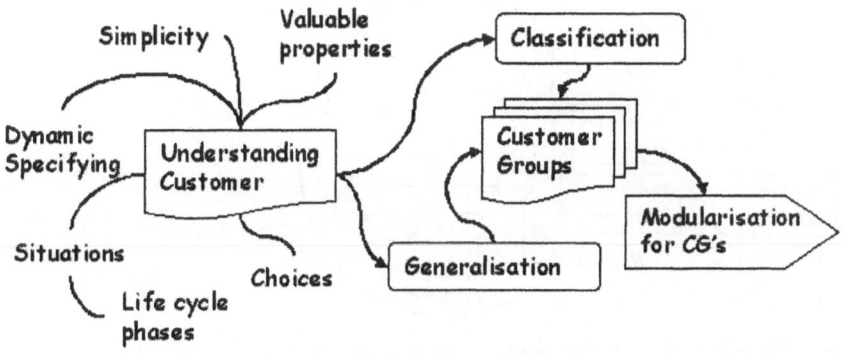

Fig.1. Understanding customer's perceptions in the sales situation

Does the opportunity to choose from a number of possibilities, options, configurations or variants make difference for a customer?

When there exists a set of customer requirements, usually an effort to categorise and group them into potential market segments or clusters is suggested (see Fig. 1). Actually it was suggested that in Modular Engineering segmenting might be dealt as a modularisation of markets. After analysis, an important task is to set the priorities to product development projects. These projects develop products, which fill the portfolio. In workshop setting the priorities were characterised as a task that "requires excellent insight".

However, the goal in configurable products is not a portfolio with a set of fixed products. Thus, it was commented that grouping customers to segments might not be valid with the development of a configurable product, because the addressed customer groups depend on the products in the sales situation. In developing platforms, module systems and configurators, the variation is deeper in the product structures and the aim is to make as general structures as possible. After all, the goal in ME should not only be the ability to develop more product variants to portfolio but also the capability to provide needed assortment with less (development, manufacturing, etc.) efforts and resources with commonality.

Similarly to markets, current product management paradigms can be segmented to at least two distinct processes. Creating portfolio or product platform and corresponding development projects (of fixed products for portfolio or modules for platform). A platform makes a framework (i.e. gives constraints and possibilities) for module development and capturing and systemising configuration knowledge.

The selection of suitable product management approach is problematic, but in both approaches "speaking the customers language" is necessary. However, there is no explicit knowledge where configurable products are at their best.

Currently we are only able to draft the business criteria for selecting the product management approach. These criteria are characterized with maturity in both technology and markets as pointed out in Fig. 2.

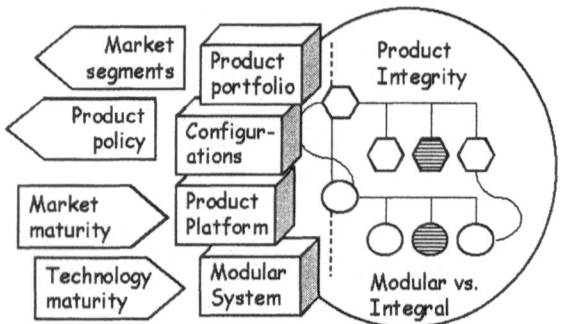

Fig. 2. The criteria for selecting product management approach

One of the business criteria is the risk related to development of ME. What are the special risks related to issues like time, cost, quality, efficiency and effectiveness, and reliability in developing and maintaining a product platform and possible configurator? If a configurator is introduced, is the underlying scheme robust enough? Does it represent the mature markets and technology? Is the platform target correct? What are the risks related to proceeding and what in delaying with platform development?

Another business issue is that it is important to understand the benefits and costs of introducing a new module type (concept) to module (configuration) system. The benefits are harvested in sales and the costs are addressed in product management. There are two conditions for developing a new concept. First, there must exist customer needs for the property the customers are willing to pay. Second, it is important to be sure that no redundancy (i.e. an already existing concept matches the emerging customer need) is created by accident.

As in all business, decisions have to be financially reasonable and justified. Product manager has to distinguish the cost of variety and relate it to perceived income. In order to relate business with customer requirements, modelling concepts (e.g. requirements, business and product entities and links) have to be developed. They have to point out the priority of requirements to development. Two types of links were suggested. One for describing the relation between variety and income and another for relating commonality and cost.

2.3
Challenges for Product Structuring and Development

In acquiring customer needs and translating them into product attributes, developing functional structures have been favoured. However, observing only functions seems to provide very limited support to product structuring. It was also stated that in product structuring research there is a lack of human elements. E.g. Theory of Technical Systems relates products' behavioural and structural elements, but does not pay attention to products as psychosocial entities. Thus, a pure technical theory is not enough for making full consideration on situations when a human being is in contact with the artefact (e.g. in sales and use).

Combinations of applications and customer segmentation were suggested as possible approaches for product structure analysis. Suggested approach for finding correct amount and target of variety included four tasks:
1) Marketing expresses simplified customer requirements (see also Fig. 1).
2) Product development creates a direct link between variety and business income (e.g. variant matrix by Volvo & Scania).
3) Management segments (modularisation of markets?) according to previous information.
4) Product experts/managers make priorities of development tasks.

Minimising complexity is becoming more important issue. Product complexity was assimilated as a product of the number of system elements and number of the element interactions. The current concepts provide means for describing only technical interactions. However, it is obvious that complexity reduction benefits are manifested when technical system interacts with human (knowledge and activities) in different life cycle phases.

It was suggested that a broader approach of product function is needed. The current ways in representing knowledge, the relation between function and physical product, and mappings between customer requirements and functions are not well defined. Are general functions needed and is selling based on products or functions, or products and processes?

Many of the current product structuring and modularisation methods deal with re-design existing products to meet the structuring goals. However, for structuring new product development there is relatively little support. Also the methods for turning an integral product to modular product platform are under research.

Structuring (or clustering) the product to modules (or chunks) should minimise the interactions between the system elements. This is usually expected to minimise the need for design co-ordination. However, one group called the modularisation effect on design co-ordination into question.

3 Conclusion

In group work several questions were asked. The most important ones are stated here. To some of them the existing theory can answer, but some remain open issues. Both the product management paradigms and the actual means for collecting and analysing customer requirements were challenged. Also drafts for early stages of product development processes were presented. Yet, no distinct paradigm neither means were generated. Instead, challenging questions, important criteria and appreciable outlines were presented and recorded for this paper.

Even Modular Products can be Unmanageable

Knut Aasland, Jarl Reitan, Detlef Blankenburg

SINTEF Industrial Management
N-7465 Trondheim, Norway
e-mails: knut.aasland@indman.sintef.no; jarl.reitan@indman.sintef.no;
detlef.blackenburg@indman.sintef.no

Abstract. Structured, modular product programs have been presented as the solution to the conflict between market and production in many companies. The value of this is not disputed, but in working with a company that has done "all the right things", new issues come up that will influence how these product programs are designed. The market aspect is crucial, because it threatens to destroy internal efficiency in the company, and handling this is an essential prerequisite for success.

1 Introduction

Within many trades and product areas there has during the last couple of decades been a development toward more and more customer-adapted products. Since customers are different, with different needs and criteria of choice, this means that it has become more and more necessary to offer a spectrum of models and variants to satisfy customers.

Variants may in some cases differ only cosmetically, but they may also have completely different specifications in functional areas. In other words, there are different degrees of "variance". What is common, is that the customer sees variants as having different qualities.

Variant diversity is a good thing, and we all know that the market desires, and to an ever greater extent demands it, but it has some production consequences that are less fortunate.

Nobody has yet been able to show that any kind of flexible production can make the total production process as efficient as series production of identical product [1].

We therefore see that companies want series production.

If we generalize, what is really desired is reuse. Not reuse of the product after it has completed its primary task, but reuse of designs, of documents, of process plans, of sales brochures and of spare part catalogues. All departments in the company will be under less strain if we consequently strive for reuse.

At the same time, certain forms of reuse – reuse of components, for example – offer other favourable effects, such as reduction of storage and administration, and thereby reduction of cost.

In other words the companies want more reuse, whereas the customers and the market want as much freedom of choice as possible. How can these opposites meet?

In order to handle the costs and incomes in a sensible manner, we need to consider the total product portfolio of the company [2]. And this means that the product to be developed must be considered not a product, but a product program, and that this product program must be made to "work" with all other products that the company manufactures.

The production of a mixture of models and variants makes efficient production difficult. The obvious solution would be to avoid variants, but the situation is that ever more demanding customers demand products that are more and more specifically tailored to their requirements. So a solution must accommodate a wide spectrum of models and variants for the market, and still allow efficient production.

The solution is the creation of production friendly product programs [1]. The key characteristic of such programs is that *"they appear as one single product from a production point of view, but as different models in the market"*. In popular terms, we can say that: "The clue is to satisfy market demands with a wide spectrum of variants, while making appear in production as only one product". One essential property of a good product program, is *reuse*. This may involve reuse of components and modules between the different variants, but may also include reuse on other levels, such as documents, tests, calculations, solution principles, and service procedures or transport solutions. The important issue is that the product program allows reuse to be exploited to its full potential.

2 The Case for Modularisation and Platform Building

Modularisation of a product increases its ability to be manufactured and assembled in steps in smaller parts. Prefabrication of modules that are especially time-consuming or difficult to produce can reduce the lead-time for finished products drastically [3, 4].

Increased use of modules in a product *family* strengthens this effect due to the fact that the modules are used in several variants. In addition to these advantages, increased use of modules opens for a great increase in the degree of reuse, something which is a key to the aforementioned additional qualities. Reuse means that *something* repeats itself in a product, and in different variants and models of a product. This *something* can be:
- Complete building units, i.e. function and assembly modules
- Groups of parts, i.e. part modules
- Solutions, i.e. scale modules
- Design elements
- User interface

What is missing in this list is reuse of components and raw materials. This comes under the term of standardization, i.e. reducing the number of components and variants by reusing chosen standard components. Modularisation has to be seen in connection with standardization. When a product has been modularised it is natural to expand the standardization to not only including standard *components*, but standard *modules* as well. Extensive use of standard modules represents a compromise between the market's demand for one-of-a-kind products and the manufacturer's demand for *one* standard product manufactured in large numbers. To Kverneland Underhaug this means to improve the relationship between product development and manufacture of potato harvesters. The modularisation work has to be connected with factory development as one of its principal targets.

So, modular products are a way to achieve product variety without production complexity.

2.1
Platform building as a way to simplify and rationalize

The term "platform" comes from the car industry. In the 20s and 30s the major American car companies launched multiple brands, to be able to appeal to a more varied public. During the 50s and 60s the brands were coordinated, so that there to a large extent came identical models under different brands [5].

The "platform", as it was then introduced, consisted of a physical building group; the frame combined with suspension and transmission. Outside the platform lay mainly the body, and to a certain degree the engines, which were brand specific.

Our definition of platform building is based on this:

A platform is a set of components and interfaces that are reused throughout the relevant part of product program, and that all in all makes up a significant part of the total product. In addition, the term platform is often used for the relevant part of the product program

The purpose of the co-ordination of the various brands was apparent. They wanted larger series. While there earlier had been Chevrolet factories and Pontiac factories, there were now platform specific factories, for instance one for the Chevrolet Cavalier and Pontiac Sunbird, and another for the Chevrolet Citation, Pontiac J2000 and other models on this platform. Platform building depends on modularisation. In many ways you can say that platform building is a successor to a modular product program, that gives even more commonality between the different models, and thus efficiency.

We can say that if there are parts of a product where the company's solutions are "good enough for anyone", and where they do not want to make special solutions for certain groups, this area will be suitable for a platform.

Platform building is a way to rationalize the product program and make it more efficient in production and logistics.

3 Kitchen Interiors as an Example

We will use a manufacturer of kitchen and bathroom interiors as an example. The company in question is Norema of Oslo, Norway. It is Norway's leading manufacturer of kitchen interiors, and they have a wide and varied product program. Their products are primarily kitchen interiors, and they deliver complete kitchens, but all electric appliances are bought in from Bosch or Electrolux. The kitchen cabinets are assembled in the factory, which distinguish them from low-cost, low profile products. Secondarily they make bathroom interiors, and also wardrobe cabinets.

The product program is of modular construction, and the ability to combine the many modules according to available space and functional requirements, as well as personal preferences, is a major point. Valach and Chal [7] present a case with similar products.

The variance in kitchen interiors is mainly of two types. The most obvious one, is the functional. This means that there are a large number of elements from which a kitchen interior can be made up, see figure 1. The elements are bench cabinets, wall cabinets and high cabinets, plus a number of auxiliary elements. Arguments for choice of these elements will have a rational element, that is, reflecting the functions one wants and the available space and other restrictions.

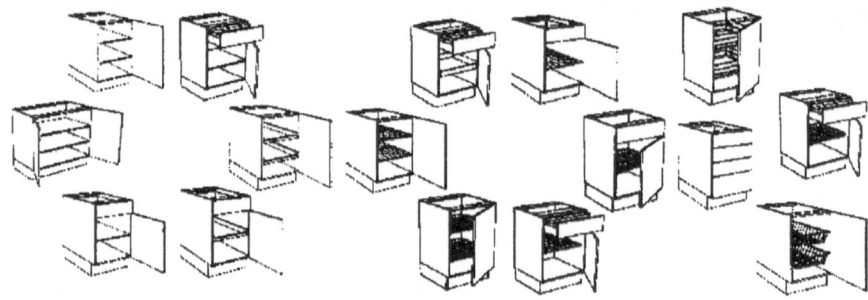

Fig.1 Functional variance

The other type of variants is aesthetic. For cabinetry, this mainly means the front material, colours and surface profiles. Whereas some more specialized – and smaller – manufacturers have only a few front types, and derive their brand image from these, larger manufacturers like Norema will usually have a wide variety of fronts and materials, and will thus offer kitchens of very different appearance.

The choice of appearance is much less rational than the choice of cabinet elements.

These two types of variance correspond to F and f as described by Hildre [9].

So we have established that most kitchen interior manufacturers see the width of their product program as a major selling point. Larger manufacturers like

Norema often look down on smaller competitors because they cannot match the choice of elements. But is it really this that determines whether or not customers choose the product of a particular manufacturer? Some information indicates that this is not the case.

Norema's products are sold solely through a chain of Norema Centers. These all have exhibitions that consist of a limited number of prototype kitchens. One experience is that many customers relate to the exhibited kitchens rather than to the available variety of elements. The argument that "I cannot have this kitchen, because mine has a different shape" is very often heard. Another effect of this is that it is difficult to sell a combination of front styles, since most centre exhibitions show "pure" style, with the same front design on all cabinets in a specific model kitchen. Fronts can, however, be freely combined, and when centers decide to show this, the effect is usually good and well received by potential customers.

4 The Problem

The case company has a planned, structured, modularised product program. And it has by all means been a success for them. The same basic concept has been manufactured for 21 years, and the flexibility allows it to be renewed and extended to accommodate new functional requirements and changing tastes.

There are, however, problems.

One of them is an *unclear brand image*. There are people who think Norema an old-fashioned and unexciting brand. On the other hand, some of their models are very "design-intensive", with high profile, modern fronts. This, however, is not easily taken by those with preconceptions about the brand.

Another, and potentially very dangerous, problem, is that their product program has developed and expanded over the years into an *unmanageable number of production variants*. Their production facilities are modern, and they support the type of product program they have, but still, the number of elements is enormous, and logistics is problematic.

So in this case, we see a company which has done most things right according to modern product program theory, and developed an efficient, modular program that appeals to customers – they are, after all, the clear market leader in Norway – but still it is not unproblematic. The issue is that customer wishes or market ideas are transformed into too wide a variety of products. This has forced a focus on the market side, with the obvious question being: "What do we really need to have in the market?"

5 What People Really Buy

This experience shows us that it is important to understand "what people really buy", that is, what they really relate to and value in the buying process. We will examine this by looking into to important sales means: the catalogue and the exhibition.

Selling by catalogue is a highly relevant issue for products like these. Many customers actively collect catalogues in order to find out where to go for more information. The catalogues have two parts; one which is meant to "lure" people into liking the kitchen in question, and one presenting the possibilities regarding elements and adaptation.

Can one sell efficiently by catalogue? The question is not a simple one. Examples show that a catalogue can ruin a possible sale, but if it can convince people that they should buy, is not so obvious. There are usually a number of steps from finding an attractive offer in a catalogue to making a decision to buy.

One particular problem is that many potential customers are not able to relate to catalogues. These people will typically not be able to transfer the information given in the model kitchens to an understanding of what it means in their kitchen. For people with such problems, a catalogue can even have the effect of scaring off people who find the products as such attractive.

The catalogue will always be a product; that is, it has a "personality" of its own. The catalogue is one of the means for the company to tell people who this company is; if it is young or old, sporty or exclusive, trendy or solid, national or international, etc, etc. The image thus created may or may not be the one that the company wants to give of itself, and that again may or may not be the most efficient way to exploit the commercial and design properties of the product.

Another important sales means for interiors, is the *in-shop exhibition*. This is where customers usually see the cabinetry and the front design in real for the first time.

One limitation of this type of marketing is that any exhibition will of course only display a limited selection of the available elements and of front treatments. In most good shops, the exhibits mimic real kitchen interiors. The design of these, combined with the build quality and professional display, is essential to the potential customer's perception of the company and the products.

In the shop, the customer will be approached by a sales representative. The quality of that experience – or lack thereof – can be a determining factor in whether or not the products will be attractive to the customer.

6 The "Extended Product" Concept

When discussing product program development, relating to "an extended product" definition is important. The customer ultimately buys more than just the physical product. People buy an experience just as much as product, that is, they relate to the experience in the sales situation, and they relate to the experience of owning the product.

For engineers, the idea of understanding the underlying arguments for decisions to buy is challenging. We believe, however, that this is a necessary precondition for making good product program decisions, so we need to dig into it.

We have found that there are many "non-functional" factors that contribute to a decision to buy. Sales psychologists say that people buy items as a tool to strengthen the image they want to give to others and themselves. This seems to be one factor, but again, only one out of many, and one which may be very important for some customers, and much less so for others.

And we should never forget that people buy according to their functional requirements also, at least those customers one might call educated.

So the product and product program must cover both of these aspects. And again the question is: "What do we really need to have?"

For some customers, being unique, or being different, is an important quality in itself. This makes it difficult to reduce variation. But can the uniqueness be limited to some minor part of the product, or even to some part of the extended product that lies outside of the actual physical product? The reason for asking this question is of course that we want simplicity and efficiency, that is reuse, similarity and standardization for production and other company internal functions.

From interviews with shop customers for Norema's products, we have found that missing functions may not be as important as the company likes to think. There are indications that "other properties" like design or the shop experience may compensate for lack of very peculiar functional modules. The way the product program is presented may be much more important than whether or not one or two variants are missing. This goes so far that we have found that even products that the customers do not consider for themselves may be of importance for their decision to buy. There are products within Norema's portfolio which are designated "talking pieces", that is, meant to cause attention and interest more than to be sold in large quantities.

7 Using this as a Guide for Designing the Product Program

The idea that customers necessarily demand an extreme degree of variation is probably not right. We need to look into the extended product concept to find ways of making a product program attractive to different customers and customer groups without having to make the physical product vary in all instances. Designing in unique elements that is supposed to appeal to special groups, could be a better way of distinguishing the product that to make extra variants.

What this means, is that the product in the limited, physical sense cannot be designed separately from the sales concept, the marketing and everything in it. Whereas this is an old statement from the early-integrated product development days, its meaning becomes much clearer when talking about product programs.

So can you create "user specific" products in the sales situation, that is, without a user specific physical product. We believe that is possible. In one interview, a kitchen interior customer (who ended up buying a competing product) referred to his product as "being the closest to a high-style designer kitchen that I can afford". Another buyer of the very same kitchen offered "simplicity, low price" as his reasons for buying. Obviously, these two customers had bought very different products, although the physical part was the same.

The issue has – to our knowledge – not been the subject of much research, and we must say that it is unresolved. It will therefore be one of the focuses of our research in the coming years.

8 Conclusions and Directions for Further Research

Understanding market mechanisms is a key to successful development of product programs. Without it, we will sooner or later end up with too wide a selection of models and variants, and thus an inefficient program. This factor has been greatly underrated in design education and in design practice.

One observation from our studies is that psychological insight is necessary if one is to understand the details that often are the difference between success and failure in a sales situation. Engineers and designers do not have the psychology background required, and inclusion of people with psychology as their education in this research is necessary. Fortunately, there are some psychologists who have "market psychology" as their business area. These people can probably contribute enormously to the understanding of the market aspect of product program development [8, 9].

One specific area that needs further study, is the possibility of replace product variance with other distinguishing elements that may convince the customers that they get a unique product. This would take seriously the slogan of "services with products in stead of products with services", that is an approach that focuses on the service aspect of what people buy in stead of the physical product.

9 References

[1] Aasland, K.: *Design for manufacturing of product programs*, in Proceedings of the ICED'97, Tampere, Finland 1997

[2] Blankenburg, D.: *Procedures for product program development*, in Proceedings of the ICED'97, Tampere, Finland 1997

[3] Blankenburg, D.: *Support concurrent engineering through product program*, in Concurrent Engineering: From product design to product marketing (proceedings of the 6th European Concurrent Engineering Conference), Erlangen, Germany 1999

[4] Tichem, M.: *Design of product families*, in Proceedings of the ICED'99, Munich, Germany 1999

[5] Tichem, M.: *Product structuring, An overview*, in Proceedings of the ICED'97, Tampere, Finland 1997

[6] Tiihonen, J., Lehtonen, T., Soininen, T., Pulkkinen, A., Sulonen, R., Riitahuhta, A.: *Modeling Configurable Product Families*, in Proceedings of the ICED'99, Munich, Germany 1999

[7] Valach, L., Chal, J.: *Handling variants and product families – A case study*, in Proceedings of the ICED'99, Munich, Germany 1999

[8] Zamirowski, E., Otto, K.: *Product portfolio architecture definition and selection*, in Proceedings of the

[9] Hildre, H. P.: *Mastering Product Variety*, 2nd WDK Workshop on Product Structuring, Delft, The Netherlands 1996

Structured Sales Process for Configuration

Seppo Suistoranta

Wärtsilä NSD Finland Oy
Stålarminkatu 45
FIN-20810 Turku
e-mail: seppo.suistoranta@wartsila-nsd.com

Abstract. Industrial markets are global and many parties from different cultures are involved in a sales situation. A particular problem is to capture customer's unexpressed needs, i.e. expectations that relate to preferences or to economic performance of the product. A structured sales process is needed. First the company has to define its business domain as a part of strategic planning. Then a concept selection process and finally a sales configurator are developed. The concept selection process is central because that is where customer's needs and expectations are transformed into a feasible solution. Structured sales process ensures that the company is able to offer solutions to its target customers based on company's own product portfolio, satisfying customer's all needs, expressed or implied. This paper presents the fundamentals of developing a structured sales process and a case study where the process is applied to a ship's propulsion machinery.

1 Introduction

The hectic business environment of today is reflected both in time and in resources assigned to projects. Sales situations vary both in terms of schedule and people. Industrial markets are global and there may be many people involved from different cultural backgrounds with a diverse way of thinking and working.

If the customer cannot express his needs or if all the relevant information is not properly understood especially two kinds of problems may arise. The first one is that the offered product does not fully satisfy customer's expectations. Later it is not easy to encourage a disappointed customer to enter a new contract. The second problem is the risk of selling something that the company's product strategy does not support which means that the company falls out of its knowledge area.

An experienced salesperson may well be aware of customer's expectations but may also lack efficient tools to handle the problem. Even if the technical requirements were clarified and a product variant that meets all these requirements created, the customer can feel disappointed.

In this paper we look for a new approach to sales process. We start from a business domain definition. It is then decomposed into three dimensions, which are customer groups, customer needs, and alternative technologies. By classifying and making concrete these dimensions we determine the area that every sales situation is based on. This makes it possible to proceed in a structured way from a sales situation into a product specification, see Fig. 1.

In complex installations, such as plants, it is typical to co-operate with several suppliers. Other suppliers' components make an extension to company's product portfolio and have to be managed together with company's own products. In these

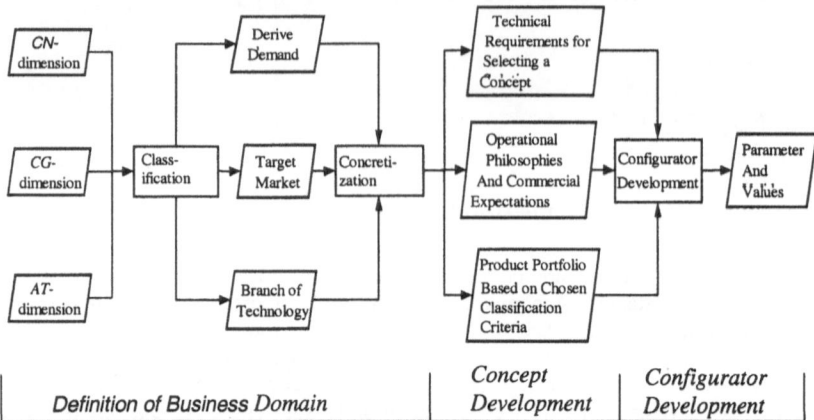

Fig. 1. Steps in developing a structured sales process

cases selection of concept is emphasized, which means composing the optimal *arrangement* of components in a way that satisfy customer's needs. Right components are then determined in a configuration process.

Sales configuration means that we create a product variant that fits to customer's needs. It is usually carried out by means of a configurator. Input data is given in parameters, which are substituted by valid values of properties. These properties are then transformed into a product description, which is a basis for a later product configuration for production and delivery process.

2 Definition of Business Domain

In order to perform a business prosperously a company has to identify the businesses it is in. A business must be viewed as a customer-satisfying process, [1]. In developing a market-based business definition we can use a concept of *business domain*, see Fig. 2.

A business domain can be defined in terms of three dimensions, which are the *customer groups* that will be served, the *customer needs* that will be met, and the *technology* that will satisfy these needs [1]. This is a general definition and needs to be processed further in order to apply it in practice. To position the company in each dimension two measures have to be carried out: classification and concretization.

2.1
Alternative Technologies

The word *technology* is used here as a generic term covering all the technical means that are capable of satisfying a set of customer needs.

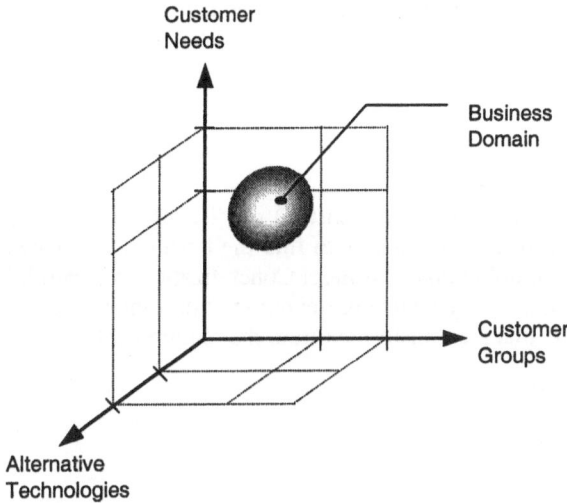

Customer Needs

Business Domain

Customer Groups

Alternative Technologies

Fig. 2. Business domain (after [1])

Generally, the technology-dimension (henceforth AT-dimension) consists of all existing technical systems. *Classification* means that these technical systems are sorted into branches (by predominant action principle, e.g. electrical systems, mechanical systems, building systems) and then grouped by some criteria into more detailed entities. This is done for strategic planning; the management has to be able to identify the branch of technology that they consider strategic for company's business.

For classification Hubka [2] presents twelve different criteria, which are:

- Function (effect)
- Action principle
- Degree of complexity
- Manufacturing similarity
- Difficulty of designing
- Production location and degree of standardization
- Design originality
- Type of production
- Degree of abstraction
- Type of operand
- Application in the technical process
- Quality

The next step is called *concretization*. It means that we try to find parameters by which the chosen branch of technology can be expressed conceptually. In reality, the AT-dimension encompasses all the physical products that the company has included in its product portfolio. A manufacturing company may include here also the knowledge area that the product development activities consider meaningful for future products.

2.2
Customer Needs

Customer needs-dimension (henceforth CN-dimension) can be interpreted as an infinite set of needs. First step in classification is to find the business area where company's target customers are doing *their business*. Concretization is a step-by-step process where we look for a more concrete presentation of customer needs. In practice we can use parameters that are transformed from the requirements list.

2.3
Customer Groups

Customer groups-dimension (henceforth CG-dimension) in the broadest sense covers all industrial markets. Classification means finding the target market where the company wants to operate. Concretization means sorting and grouping the customers in the target market by certain criteria of operational similarity. In broad terms, the customers whose way of performing business operations are equal or close to each other belong to the same customer group. The highest level of concretization is expressed in customer-related requirements for determining operational philosophies and commercial expectations.

Mass customization aims at treating all customers individually and making products that are tailored for each specific customer. This is not contradictory with customer grouping and further it can be justified for practical reasons. Customer group is a term that covers those individual customers who have one or more element in common. We can group customers by geographical area, by sales regions, or by financial potential; they all have a practical meaning in marketing activities.

2.4
System of AT, CN, and CG

The three dimensions introduced above can also be interpreted in another way, which emphasizes independence of these elements. Using the terminology from Hubka [2], we present the following system model:

"Alternative technology" is a name for a transformation system. *"Customer needs"* is an operand, which is transformed in the *"AT"* from unsatisfied state into satisfied state. *"Customer group"* (as a common term) is an operator that operates the transformation system. This interpretation implies that the customer really operates (uses) the transformation system (product) and his satisfaction depends on its outcome.

3 Process of Concept Selection

To work a solution concept into a configuration requires knowledge of the customer and of his business. Salesperson's experience is then necessary for evaluating the concepts against customer's expectations.

For this purpose a structured process is needed for merging our knowledge of the customer, his needs, and our product portfolio. We are looking for a process whose output is a valid solution concept, so that: 1) it can be compiled from company's product portfolio and its extensions, 2) it can be offered to company's target customer, and 3) it satisfies needs of this customer.

3.1
Technical Requirements

Technical requirements originate from the customer's needs and can be divided into technology-specific, application-specific, performance-specific, and installation-specific groups. When these requirements are fulfilled the product performs the desired functions and fits to the technical environment as the customer has designated.

In complex installations (industrial plants, ship's engine rooms, etc.) the product consists of a set of machines, devices, and apparatus that work together. These components can be configurable products or built as one-of-a-kind. In theory, there are an infinite number of valid arrangements of components. It is also obvious that there are several feasible solution concepts.

Statement 1: There are several feasible solutions that satisfy customer's requirements.

3.2
Customer's Expectations

In a sales situation it may be difficult to capture customer's unexpressed needs. We call them collectively *customer's expectations*, covering both economic and other matters. We find it purposeful to introduce this term instead of using *latent needs*, which relate more or less to development of new product concepts. "Latent needs are those that many customers recognise as important in a final product but do not or are not able to articulate in advance", states Ulrich and Eppinger [3]. Customer's expectations relate to an existing product or a set of products and to their measurable outcome. In broad terms, they reflect the value that the product delivers to the customer.

It is not simple to measure the value delivered to the customer and theories of this issue go beyond the scope of this paper. In this context we assume that the customer is primarily interested in benefits that the product carries to his business. However, in reality every customer has a slightly different view to these benefits, depending on their business and operational philosophies.

Sales configuration is a process that helps in specifying the product according to the needs of customers. It makes a structured basis for the production and the whole delivery process. The sales configuration process utilizes parameters that are often called *properties*. They may have a point value (an option) or a specified value in a valid interval. In a sales situation these parameters are often discussed and juxtaposed with the external properties, especially the functionally determined properties of the product. In this way they make a common language between the parties, [5].

Hubka [4] states: "The Theory of Properties explains that each technical system carries all the kinds of properties, especially those that make it suitable for a purpose. We distinguish between external and internal properties. The realized technical system owns (possesses) therefore all properties (features), whether they are deliberately planned, or have not been considered".

A configuration process means that with a limited set of properties (configuration parameters) we create a product that carries all possible properties associated with it. This implies that N customers whose *functional* requirements are alike may arrive at same set of input parameters leading to equal configurations. However, due to diverse businesses these customers may have a totally opposite view to the product. This is significant, since in many cases the customer is making an investment with a precise payback time and may have a lot of expectations that relate to the economic performance of the product.

Statement 2: Between technically equal solutions the customer chooses the one that delivers most value to him.

This leads us to a conclusion that in a sales process we should emphasize the concept selection phase, because it is there that the customer's expectations are taken into account. Later in the configuration phase it is impossible to correct any fundamental errors or at least a lot of time and money is lost.

The task is now to identify parameters by which to represent customer's expectations formally. In the concept selection phase we utilize these parameters by giving valid values to them and obtain a preference profile of a specific customer, which is then used as a standard for evaluating different conceptual solutions. In this task the tacit knowledge of experienced salespersons is needed. It does not necessarily call for electronic data processing; rather it is question of a systematic way of working.

3.3
Product Portfolio

It is obvious that a company tries to sell products from its own portfolio. However, especially in complex installations it is quite typical that components from other manufacturers are used in parallel. It is beneficial for all parties to work together and try to develop a common way of working.

If the solution concepts are largely based on components purchased from other suppliers it is recommended to extend the process to cover also them.

4 Configurator Development

The creation of a particular product variant is a configuration task. A crucial element in configuration is a precise fitting of the products to the needs of customers [6].

A sales configurator generates a technically feasible configuration as long as the input parameters have been acceptably given. This means that there may be several solutions that satisfy customer's technical requirements.

A much-referenced definition of a configuration task is as follows ([6] after Mittal):

Given:

(A) A fixed, predefined set of components, where a component is a set of properties, ports for connecting it to other components, constraints at each port that describe the components that can be connected at that port, and other structural constraints, (B) some description of the desired configuration; and (C) possibly some criteria for making optimal selections.

Build:

One or more configurations that satisfy all the requirements, where a configuration is a set of components and a description of the connections between the components in the set, or, detect inconsistencies in the composition.

Mittal's definition implies that in configuration it is actually question of managing *properties,* encompassing also the criteria for *making optimal selections.* The "optimal selection" can be understood both in terms of technical goodness and of suitability from customer's point of view.

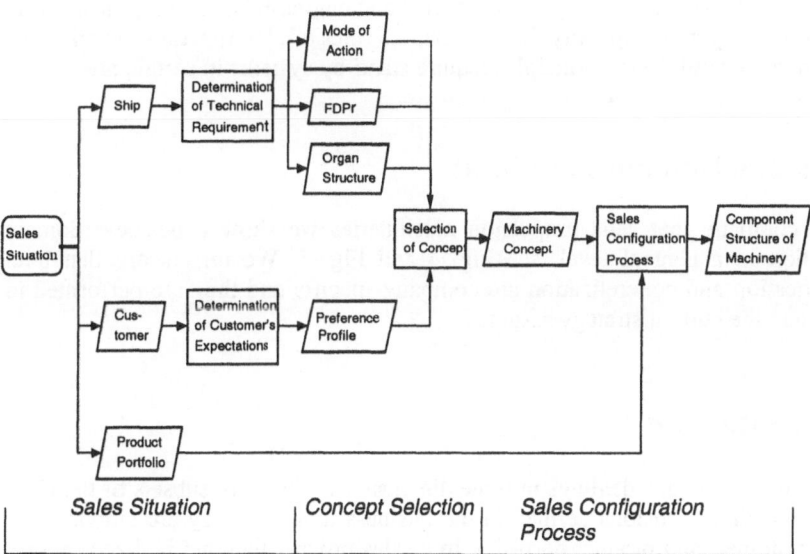

Fig. 3. Structured sales process of a ship's propulsion machinery

We state that the objective in configuration is not only to find a feasible solution to perform a set of desired functions but more like to realize those properties, which contribute to success of customer's business. One essential objective in configurator development is to enhance the groups of configuration parameters to cover also customer-related requirements.

5 Structured Sales Process of a Propulsion Machinery

5.1
Propulsion Machinery

Propulsion machinery is an energy generating plant, which can be categorized with complexity of highest level. It is a set of individual machines and equipment that work together in a controlled way. The main function of propulsion machinery is to generate propulsion power for a ship. In principle a simple organ structure that consists of a prime mover (engine) and a shaft line (coupling, reduction gear, support bearings, and propeller) can represent it. Typically propulsion machinery has also desired secondary functions, such as electric power generation (shaft generator) and heat generation (waste heat recovery system).

Complexity of propulsion machinery is due to the requirement of self-sufficiency of the entirety. To maintain continuous operation several auxiliaries and sub-systems are needed: storage, treatment, and delivery systems for fuel oil and lubrication oil, compressed air system, cooling water systems, etc. Multi-engine installations have more prime movers and engine-related auxiliaries, giving rise even to higher complexity. In addition, the rules of classification societies that govern the overall design principles require stand-by systems in certain areas.

5.2
Business Domain Definition

For a company that sells propulsion machineries we show a business domain definition in a general level, see Fig. 4 and Fig. 5. We underscore that both classification and concretization are company-specific and they are performed to the extent the current strategy assumes.

5.3
Sales Situation

Sales situation means dealings in three dimensions, which are subsets of the CN-, CG-, and AT-dimensions defined in the business domain. They are embodied in ship, customer, and product portfolio. In a sales process they get realized in a set of parallel activities that finally end up at a component structure of propulsion machinery, see Fig. 3.

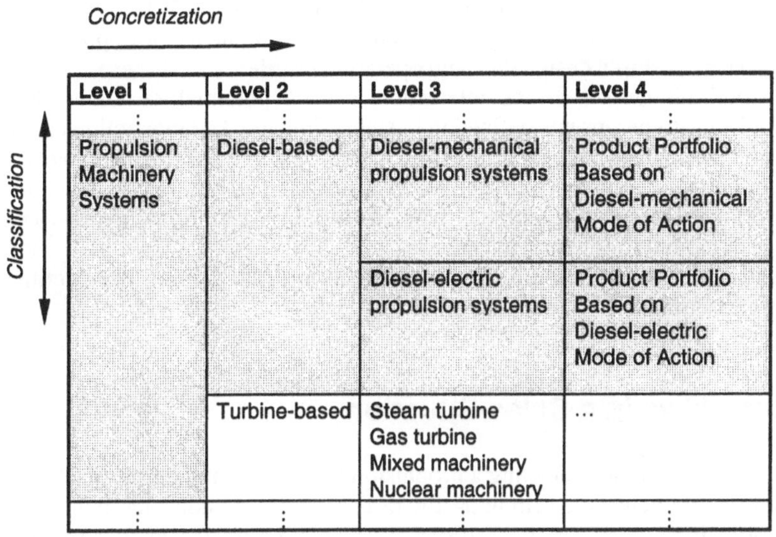

Concretization

Level 1	Level 2	Level 3	Level 4
⋮	⋮	⋮	⋮
Propulsion Machinery Systems	Diesel-based	Diesel-mechanical propulsion systems	Product Portfolio Based on Diesel-mechanical Mode of Action
		Diesel-electric propulsion systems	Product Portfolio Based on Diesel-electric Mode of Action
	Turbine-based	Steam turbine Gas turbine Mixed machinery Nuclear machinery	...
⋮	⋮	⋮	⋮

Classification (vertical axis label)

Fig. 4. Classified and concretized AT-dimension

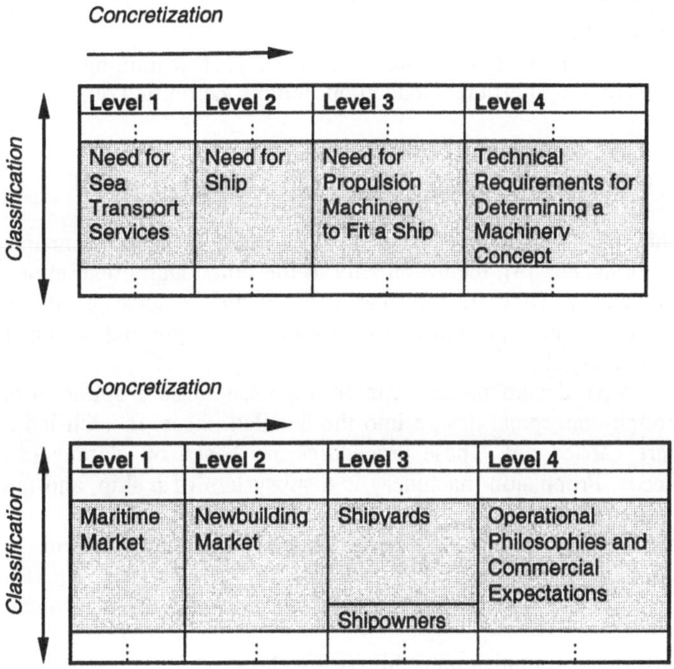

Concretization

Level 1	Level 2	Level 3	Level 4
⋮	⋮	⋮	⋮
Need for Sea Transport Services	Need for Ship	Need for Propulsion Machinery to Fit a Ship	Technical Requirements for Determining a Machinery Concept
⋮	⋮	⋮	⋮

Classification (vertical axis label)

Concretization

Level 1	Level 2	Level 3	Level 4
⋮	⋮	⋮	⋮
Maritime Market	Newbuilding Market	Shipyards	Operational Philosophies and Commercial Expectations
		Shipowners	
⋮	⋮	⋮	⋮

Classification (vertical axis label)

Fig. 5. Classified and concretized CN- and CG-dimension

Action in the first dimension starts with clarifying all relevant information of the ship in order to determine the technical requirements for propulsion machinery, such as ship's design data and operating profile.

The second dimension means interaction with a potential customer, who belongs to the customer groups defined in business domain. The interaction can be a negotiation or a call for bid, etc. In reality, there may be several parties involved but for simplicity we term them collectively a customer. The first step is to acquire customer-related information, such as type of shipping operations, experiences, financial background for the project, etc.

The third dimension is built up from the current portfolio that has been defined in the business domain.

5.4
Different Views in a Sales Situation

Salesperson's view: When selecting a propulsion system it is important to take the operational requirements and sailing profile of the vessel into consideration. If the sailing profile is complex or there are space limitations in the engine room, a whole spectrum of propulsion arrangements is possible, [7].

For selecting mode of action several factors have to be assessed [8]:
1. Influence on selling price
2. Transmission losses
3. Flexibility requirement, i.e. the installed prime mover capacity may be used for different purposes in different situations
4. Machinery uniformity, which affects spare parts logistics, crew training etc.
5. Requirement for maintaining efficient load at optimum specific fuel oil consumption
6. Location of main engines to optimize ship arrangement
7. Requirement for redundancy

The powering of the ship, which depends in the first place on ship's tonnage, speed, ambient conditions, etc. [8], relates directly to the functionally determined properties of the machinery. Other factors that affect them are location of the engine room and its general arrangement that may limit the weight and principal dimensions of the machinery [9].

Shipyard's view: Ship's design process can be represented as a spiral from requirement list through conceptual design into the detailed phase. In each lap a set of procedures are carried out. These procedures are based on shipowner's requirements and needs. Propulsion machinery is a subsystem of a ship, and the following steps are taken [9]:
1. Propulsion machinery that satisfies the performance-specific requirements is chosen
2. Space requirements of the chosen machinery is checked against the engine room
3. Auxiliary systems for the propulsion machinery are chosen
4. Auxiliary systems and the cargo handling equipment for the ship are chosen
5. Power and heat generation capacity is checked

6. Fuel consumption of the chosen propulsion machinery as well as auxiliary machinery is calculated
7. Weight and center of gravity of the machinery and fuel is calculated
8. Environmental impact (vibration and noise excitations, emissions) is checked against the requirements
9. Required number of operating people is checked
10. Price of the machinery is checked and compared with a detailed cost calculation
11. Modifications to ship's main dimensions and/or to general arrangement are made, if necessary

Apart from these, the shipowner may have wishes and viewpoints that are difficult to express in numbers, such as requirements based on routines and operational experience.

Shipowner's view: The shipowner views the propulsion machinery as an investment. He is interested in machinery's purchase price and suitability for duties. Operational properties of the machinery – such as availability, reliability, operability, serviceability, etc. – interact with economic performance. In broad terms, two issues capture shipowner's view: life cycle costs and quality.

5.5
Determination of Requirements and Customer's Expectations

In determining *technical requirements* we process the received information aiming at a structured set of data that is necessary for a selection of propulsion machinery concept. The data can be categorized into technology-specific, application-specific, performance-specific, and installation-specific requirements.

Customer's expectations are best captured by *economic performance* associated with the vessel. A ship is an investment and the shipowner expects a reasonable profit in providing sea transport services with it. The shipowner is particularly working with three cashflow items: 1) revenue received from chartering or operating the ship, 2) cost of running the ship, and 3) method of financing business [10].

Stopford [10] states: "There are several ways a shipowner can earn revenue, each of which brings a different distribution of risk between the shipowner and the charterer and a different apportionment of costs. Each of the revenue arrangements deals with these items differently." Three typical revenue arrangements are voyage charter, time charter, and bareboat charter.

In the shipping industry there is no internationally accepted standard cost classification. Stopford [10] Presents five cost categories: 1) operating costs, 2) periodic maintenance costs, 3) voyage costs, 4) capital costs, and 5) cargo handling costs. Each of these categories can further be broken down into more detailed sets of costs, e.g. voyage costs encompasses heavy fuel oil, diesel oil, and port costs.

Because the propulsion machinery is a subsystem of the ship, we propose that it is both a *revenue-driver* and a *cost-driver* to shipowner's business. By making a rough simplification we may assume that the shipowner is interested in how the selected propulsion machinery can contribute to shipping revenues or save in running costs.

The items listed in the cost classification may be charged to different parties (shipowner, charterer), depending on the revenue arrangement.

5.6
Selection of Machinery Concept

Selection of machinery concept means determining three principal issues: 1) the mode of action, 2) the organ structure, 3) functionally determined properties.

Hubka [4] characterizes *mode of action* as follows: "Effects can be achieved by different natural processes... The same effect (output function) can be realized in several ways, termed the mode of action of technical systems...Each mode of action demands a certain structure, which guarantees these internal trans-formations. According to organization principle (construction principle) several kinds of construction emerge for the structure, which must all deliver the same output."

In a case of propulsion machinery we distinguish the mode of action by the principle of power transmission, which can be *diesel-mechanical* or *diesel-electric*. The selection is mainly based on application-specific requirements.

Organ structure defines organs and their relationships. Organs are: prime movers, shaft lines, auxiliary systems (to be precise, each of these organs is in reality an *organism*).

Our task is to find a satisfactory organ structure for propulsion machinery. We define that the organ structure is satisfactory, if the organs can be mapped to components (more precisely: machines, equipment, device) that have valid performance ratings and that together make a compatible system.

Functionally determined properties belong to a class of external properties and they include e.g. performance ratings (speed, power, functional dimensions, size), suitability for duties and environment, and secondary outputs [4]. At this stage it is not necessary to determine quantitative values for these properties but to assess which type of engine (slow speed, medium speed, high speed) and which magnitude of power output is purposeful for the case.

5.7
Configuration

In the concept selection process we have determined the technology-specific, application-specific, performance-specific, and installation-specific requirements for the prime movers. Mode of action, satisfactory organ structure, and functionally determined properties are merged into a solution concept, which is then evaluated against customer's preference profile.

In a configuration process a component structure is generated. It specifies each component by type, ratings, manufacturer and other relevant information and in its entirety makes a representation of compatible subsystems and realizable propulsion machinery. This means in practice that we have determined the mode of action (diesel-mechanic or diesel-electric), the engine type (slow speed, medium speed, high speed), ratings (power output, speed), weight and principal dimensions and interfaces to the engine room systems (pipe outlets and inlets, location of turbochargers etc).

6 Conclusion

In this paper we studied a general model for a structured sales process to better take into consideration customer's needs. The first step is to define the business domain, which captures three strategic components of business, i.e. alternative technologies, customer needs, and customer groups.

The second step is to develop a concept selection process. This is where customer's expectations (or unexpressed needs) are taken into account. Tacit knowledge of experienced salespersons is needed for determining a standard, which in every sales situation is transformed into a customer-specific preference profile. It is then used for evaluating the selected solution concept against customer's expectations.

Third step is configurator development. Typically the objective in configuration is to determine the best possible product variant that fits to customer's needs. This paper called for a more customer-oriented objective: configuration for contributing to customer's business.

Finally, in this paper we presented a structured sales process of a ship's propulsion machinery as a case study. Propulsion machinery is a complex installation, which is composed of several components (machines, devices, instruments, etc.). The solution concept is selected by three main factors: mode of action, satisfactory organ structure, and functionally determined properties. The concept is then evaluated using customer's preference profile. In this way we arrive at a solution that has the best economic performance for customer's business.

7 References

[1] Kotler, Philip: *"Marketing Management. Analysis, Planning, Implementation, And Control."* Seventh Edition. Prentice-Hall International Editions. Eaglewood Cliffs, New Jersey.

[2] Hubka, Vladimir – Eder, Ernst E.: *Theory of Technical Systems. A Total Concept Theory for Engineering Design.* Springer-Verlag. Berlin Heidelberg.

[3] Ulrich, Karl T. – Eppinger, Steven D.: *"Product Design and Development".* McGraw-Hill International Editions.1995.

[4] Hubka, Vladimir – Eder, Ernst E.: *"Design Science. Introduction to the Needs, Scope, And Organization of Engineering Design Knowledge".* Springer-Verlag. Berlin-Heidelberg-New York. 1996.

[5] Suistoranta, Seppo – Riitahuhta, Asko: *"Product Structuring in a Globally Operating Company".* Int. Conf. on Engineering Design, ICED 99, August 24-26, 1999, Munich, Germany.

[6]   Riitahuhta, Asko – Andreasen, Mogens Myrup: *"Configuration by Modularisation"*.
      Proceedings of the NordDesign '98, Stockholm 26.-28.8.1998, p. 167-176.
[7]   Wärtsilä NSD Corporation: *"A Guide for Selecting Marine Propulsion Systems"*.
      Zürich, 1999. (Not published).
[8]   Henriksson, Torbjörn: *"Machinery for RoPax and Cruise Ships"*. Wärtsilä NSD
      Finland Oy. Application Technology/Marine Engines. October 4, 1999.
[9]   Häkkinen, Pentti: *"Laivan koneistot"*. Teknillinen korkeakoulu, Konetekniikan osasto,
      Laivalaboratorio. M-179. Otaniemi 1993. ISBN 951-22-1780-5. *("Ship's Machin-
      eries." Helsinki University of Technology, Faculty of Mechanical Engineering, Ship
      Laboratory. Report M-179. In Finnish.)*
[10] Stopford, Martin: *"Maritime Economics"*. Second Edition. Routledge, London.
      Padstow, Great Britain.

Visualizing Interdependencies in Technical Design

B.J. Oosterman, G.J.C. Gaalman, and F.P.J. Kuijpers

Faculty of Management and Organisation
University of Groningen
9700 AV Groningen, The Netherlands
e-mail: b.j.oosterman@bdk.rug.nl

Abstract. The control and understanding of complex design processes is an important issue in practice and theory. Recent research shows that documentation and analysis of technical interactions between design tasks (using Design Structure Matrices) gives a thorough understanding of the underlying complex structure of design processes. Generally, these studies are based on an existing set of design tasks and interactions and claim that understanding a design structure is particularly helpful to improve (future) design processes. However, we argue that this claim can not be demonstrated because current studies do not document technical cause and nature of interactions. This hampers an objective understanding of design structures of future design processes. In this paper we propose a taxonomy of technical interactions based on product architecture in order to obtain descriptive clarity and better understanding of cause and nature of technical interactions. This will help to better focus on coordination where it is most needed. Further it gives structural insight how to make significant changes in product architecture in order to improve future management and organization of technical product development.

1 Introduction

The last decades, companies have been forced to improve and speed up their product innovation because of changing technology, globalization, and more demanding customers [1]. The control and understanding of complex development projects is therefore an important issue in practice and theory.

Design processes involve many aspects varying from idea generation to production and commerce, and are usually split up into many interacting design tasks. In fact, many studies about improvement of development processes focus on effective coordination between all development activities. For example, general design procedures have been developed to aim for logical and transparent design processes across all designers [2, 3]. Further, Concurrent Engineering (CE) achieves better coordination by organizational actions like multi-functional teams, lateral roles, overlapping of phases, and applications of analytical tools such as DFX, QFD, and FMEA. The use of these integrating approaches leads to improved product quality, lower manufacturing costs, and a shorter Time-to-Market [4].

Though this stream of research presents very useful insights, it does not focus on what makes development projects complex and where most coordination effort is needed. The measuring of complexity has been an important issue in science for many years and many different definitions of the construct exist [5, 6]. For example, Simon [7] defines a complex system roughly as one made up by a large number of elements that interact in a non-simple way. According to most definitions interdependence between elements of a system is seen as an important factor of complexity. Many researchers have studied the causes and effects of interactions between elements. Simon argues that most systems are nearly decomposable, which means that not all interactions between sub-systems are equally strong. Probably because of this property, strengths of interactions have been analyzed from many perspectives. Many scholars have studied interactions between individuals, among sub-units, among organizations, between hierarchical levels, and between tasks [8]. For many years organization scientists have studied complex organizations and propose several organizational structures to handle complexity [9, 10]. Here, a central theme is to co-ordinate the distributed work across people effectively in order to perform the main task of an organization well. To minimize and simplify coordination effort decomposition of a system in rather independent sub-systems is a powerful strategy [9]. Once a system is decomposed, remaining interactions between sub-systems should be well understood and effectively coordinated for the overall system to perform well.

In sum, for most complex systems decomposition decisions strongly influence the need for coordination, and knowledge of remaining interactions between sub-systems gives insight in where, and how to co-ordinate. In the next sections we will discuss how these ideas are used in complex product development projects to address the following questions: How to analyze interactions in the design of technical products? What is the role of product architecture in this, and how can we use this architecture in order to understand and improve the management of design processes?

This paper is organized as follows, section 2 surveys DSM models and the role of interactions in design processes. Section 3 present some critical remarks about the documentation of tasks and definition of interactions in current DSM studies. Section 4 describes product architecture and the nature of technical interactions. Next, section 5 presents the basis for three types of technical interactions between product modules related to product architecture. Section 6 visualizes these interactions and discusses the implications for coordination between design tasks and options for improvement. Finally, section 7 summarizes the most important findings.

2 DSM Models

Von Hippel [11] suggested that partitioning a design process in tasks and interactions is a crucial decision that has a high impact on design process efficiency. Recently, some scholars model and analyze these tasks and interactions, mostly based on Design Structure Matrices (DSM) [12]. Usually a set of design tasks and interactions is obtained by interviewing several project team members. The design process is then modeled as a set of information processing design tasks and necessary exchange of information between the tasks. The design tasks, location and direction of interactions are documented in a DSM matrix such as depicted in Fig. 1. More precisely an element (i, j) of the matrix indicates whether task *i* requires the output of task *j* or not. The whole column of task D, for instance, shows that task E, F and G need input from D to be performed.

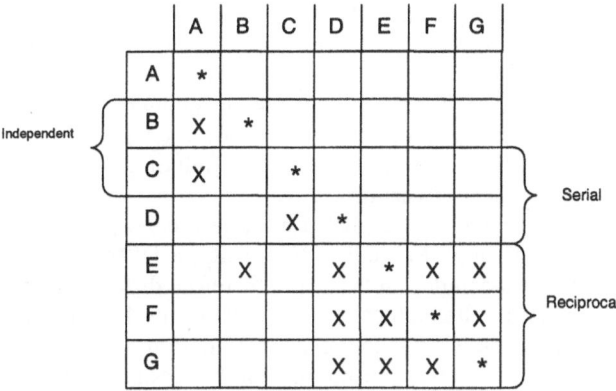

Fig. 1. Modeling tasks and interactions in a DSM

The DSM method is a very robust technique that can be applied to many problems to analyze the underlying design puzzle, and give both mathematical and deep practical understanding.

Identification and mathematical analysis of (groups of) design tasks for which the output of each becomes the input for the other (reciprocal interdependency) gives a formal understanding of the need for design iterations (rework). For example, Smith and Eppinger [13] formulated a DSM to analyze a design process of an automotive brake system. It worked out that analysis of complex, reciprocal interactions and corresponding iterations gave a good estimation of duration and possible variation of design process lead-times. Along with this, familiar studies focus on concurrency, amount of overlap, costs, and clustering and scheduling of strongly coupled tasks [14, 15, 13, 16].

The DSM model does not explicitly include methods to co-ordinate, but mechanisms can be proposed to co-ordinate the documented interactions in order to reduce or speed up iterations in design. For example, similar to Thompson's theory a team of designers can be grouped around a cluster of reciprocal tasks. In this way strong interactions can easily be adjusted mutually, while other coordination modes take care for the interactions with other tasks. For a redesign project of a small block V-8 automotive engine a DSM analysis led to a grouping of 24 design teams into four system teams to improve overall coordination [17]. DSM models give common understanding of interactions such that formal coordination modes can be proposed instead of trusting on informal coordination and expecting continuous overview of designers.

In fact, most DSM studies show very useful applications and conclude that the concept lays an eminent basis for improvement of future design processes. In the next section we will especially focus on this claim and have a more critical view at the DSM studies.

3 Exploring Tasks and Interactions

Just like many classical design theories the DSM studies (implicitly) base the partitioning and coordination decisions on a well-defined set of tasks and interactions at a given moment. To contribute to improvement of design processes, the definitions of tasks and interactions should be clear and unambiguous. Further, the analysis of tasks and interactions should ideally be valid for a longer period of time. Therefore we will explore the meaning of the used definition of tasks, the definition of interactions, and the role of dynamics.

First, similar to most organizational studies that use the task concept [18], no adequate definition of tasks exists in DSM studies. Especially, a DSM does not model hierarchical differences between design tasks and therefore different levels of detail can not be recognized. Hierarchical differences occur because for example, a design process can be modeled as one main task 'design product' that can be split up into several sub-tasks such as 'design module 1', 'design module 2' etc. Next, a subtask can be further decomposed such that a hierarchical structure can be formulated such as depicted in Fig. 2. However, in a DSM each documented task is arranged similarly, and multiple hierarchical levels of design tasks can exist next to each other (such as depicted in Fig. 2). As a consequence, this complicates a self-evident analysis because this critically depends on the modeling of design tasks at the same level of detail. For instance, imagine the difference in structure if we model one aggregate task instead of the reciprocal coupled subtasks E, F, G depicted in Fig. 2.

Second, interactions are defined as exchange of information, thus only concerning location and direction of interaction. It is clear which tasks do need to exchange information (location), but the cause and nature of this interaction can not be obtained. As a consequence, some methodological difficulties may occur, because gathering data of interactions is sensitive for the focus, perception, and background of project members at a certain moment [19]. Interdependencies have a multidimensional character and include aspects that vary from psychological to pure technical interactions [20, 21, 8]. Some studies restrict to technical interaction, but this still leaves open whether all interested parties have a common interpretation or not.

Third, as time is concerned the question is whether an analysis will stay valid when an organization shifts towards another set of interacting tasks. To our best knowledge most published DSM studies are retrospective and describe how the method could have worked [16]. However, the validity of a partitioning scheme and organizational actions may be influenced by the dynamics of interactions over time. Most researchers address this problem and constrain the applicability of the DSM models to redesign projects of existing products. Apparently, the structures of design processes of these products seem to have a high similarity. However, it is left to the user to judge what aspects of the analysis can be generalized to future projects and which are probably subject to change. Moreover, the before mentioned lack of clarity about the causes of necessary information exchange also hampers insight in how and where we eventually can consider a technical redesign in order to decrease coordination effort.

Hierarchy of design tasks

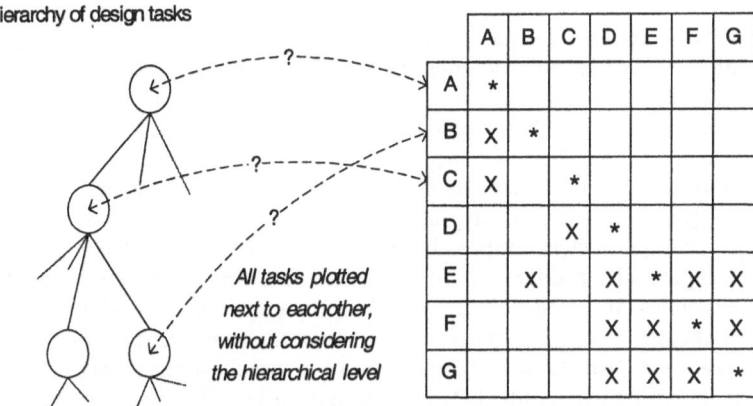

Fig. 2. Hierarchy and DSM

In sum, if we focus on the role of DSM models to improve future design processes we conclude that a good indication of hierarchy, and insight in nature and cause of technical interactions is required. This makes an analysis more self-evident and also indicates how interactions may change over time or how an organization eventually can influence its set of technical interactions.

In the next section we will explore why results of current DSM studies are particularly helpful for products that share the same product architecture. Here, we will consider product architecture and its relation with interactions between design tasks and corresponding co-ordination effort.

4 Product Architecture and Nature of Interactions

We define product architecture in informal terms as how distinct physical modules of a product technically interact in order to obtain a functioning of the whole. The type of product architecture has major influence on many aspects in the organization [22]. Therefore, the choice of a more or less modular product architecture needs a careful consideration of many factors such as production, logistics, strategy, external suppliers and customers, environmental constraints, technology, performance, costs, and product development processes [23, 24, 22, 2, 25]. Obviously, product architecture is embedded in the total organization and the type of architecture is not likely to change frequently [26]. Consequently, we restrict this paper to redesigns of existing products that improve the function or the technical concept, but do not change the type of product architecture.

In case of an existing totally modular product architecture, physical modules can be changed independently without changing the specifications of technical interactions. Hence, modular product architectures serve as a stable intermediate form from which new products can evolve very fast. Many successful families of products, such as the well-known Sony Walkman [27], were incremental innovations based on stable product architectures. Because the specifications of technical interactions between modules do not change within a design process, relatively little information needs to be exchanged between design tasks during the design process and relatively little coordination effort is necessary.

Contrary, underlying technology and the need for fashionable styling or for low unit costs may not permit a modular design and a product should be designed more integral [28]. For a totally integral architecture a small improvement of the functioning of the product is not possible without effecting all 'physical chunks' and thus changing all specifications of technical interactions between chunks. Think for examples of a mini camera, small laptops, or integrated chips, where even a small change in size of a component can be of major influence [28]. Design processes based on an integral architecture have a more complex structure and need much coordination effort between design tasks. Because for integral architectures the term module may be misleading, we will use the term physical chunk in the following. More clearly, a physical chunk is a separate part or assembly of a physical product.

In sum, we see that the type of product architecture gives us a lot of insight in the nature and role of interactions. For a design process it determines the amount of change in specifications of technical interactions between modules [26] that in turn relates to the need for coordination between design tasks. Further, the product architecture influences the amount of similarity between technical interactions of different derivative products.

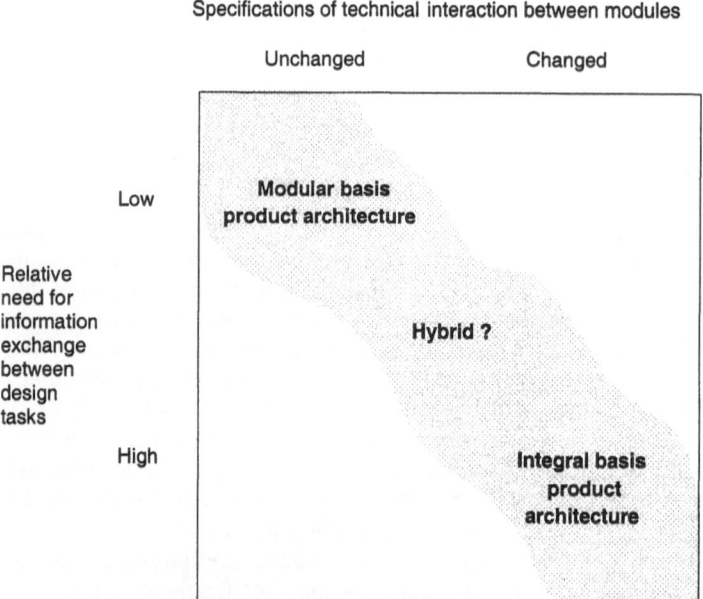

Fig. 3. Effect of innovations that reinforce a function, per type of product architecture

However, most products have architectures that lie somewhere between modular and integral. As a consequence, the relative need for coordination and the amount of change in specifications depends on the type of interactions between pairs of modules (see hybrid in Fig. 3). So we need a more detailed analysis of technical interactions between physical chunks and its relation to the structure of design tasks. In the next section we will define three types of technical interactions that relate to product architecture. The relation between technical interactions between physical chunks and the structure of design tasks will be presented in the subsequent section.

5 Typology of Technical Interactions

To explore types of technical interactions we separate the functional and physical part of an existing product [29] such as depicted in Fig. 4. Functions in the functional domain of a product describe what a product has to do in order to fulfil client needs (costs, performance requirements etc.). The physical chunks in the physical domain describe how the functions of a product are established by technical solutions that are fulfilled by (detailed) physical characteristics. These physical characteristics are subject to many constraints such as ease of production, ease of assembly, costs of the bill of material, test requirements, available space etc. Both functions and physical chunks can be described in a hierarchy, where the understanding of details of a domain increases by moving down the hierarchy,

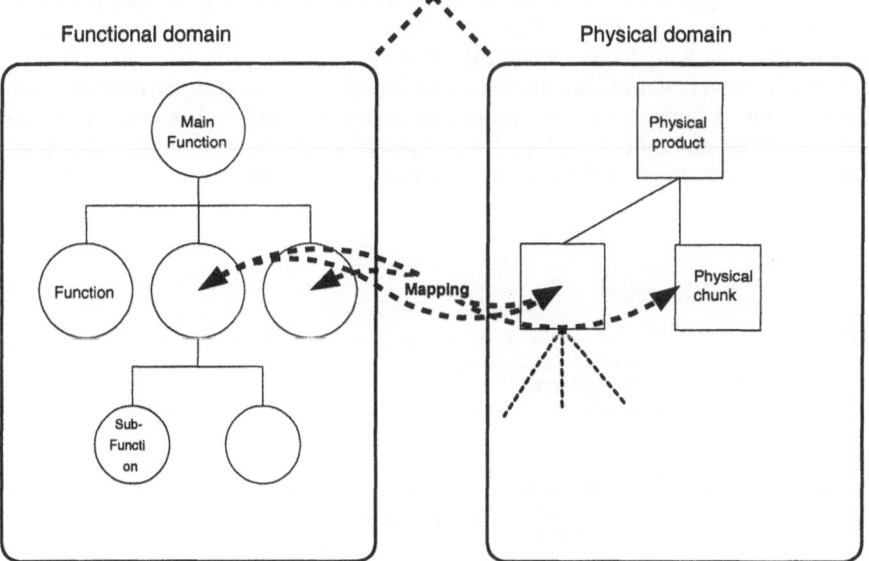

Fig. 4. Separation of the functional and the physical domain of a product.

while in moving up a better understanding of its significance is obtained. Further, the mapping between the two domains specifies which physical chuck(s) perform(s) which function(s). Now we have presented a functional domain, a physical domain and the mapping between the domains we visit the definition of product architecture in more detail. Ulrich [29] describes product architecture in formal terms as:

'The arrangement of functional elements.'
'The mapping from functional elements to physical components '(chunks).
'The specification of interfaces among interacting physical components (chucks).

Because we aim to formulate technical interactions that relate to product architecture, we propose that physical chunks may have interactions because of (1) interactions between functions, (2) a non 1:1 mapping from functions to physical chunks, and (3) because the chunks physically have to be attached to each other.

5.1
Interactions between functions

No general formal language exists to describe and relate functions but globally a function is an abstract formulation of what a product has to do, without specifying any particular solution [2]. Functions of a product can be modeled in a so-called function structure that describes functions in a hierarchy of functions, sub functions, and sub-sub functions, etc. At a certain level of detail, most functions can be described in terms of input and output of energy, material, or information [30]. Accordingly, we say two functions interact if the output of the one is needed as input for the other. For example, two functions of an electric shaver 'supply power' and 'cut hair' have a technical interaction because they have to exchange energy. However, not all functions can be linked by input- output relationships. For example, a function such as 'protect against water' will be hard to describe in terms of input and output. In Fig. 4 we depicted a schematic functions structure, where we do not consider the relation with the environment.

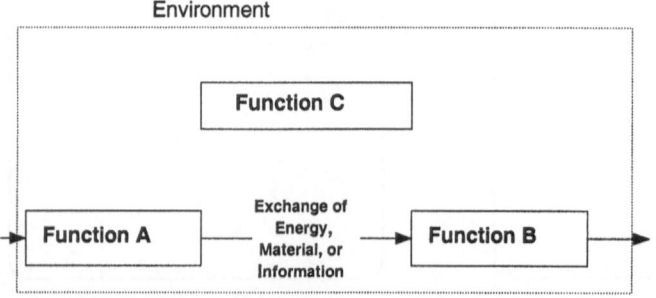

Fig. 5. Function structure of a product

5.2
Mapping from functions to physical chunks

Physical chunks implement the functions of the product. Now, suppose two physical chunks each perform a function and these functions need the right specification of input and output of energy, material, or information from each other. This is plotted in Fig. 6 for functions A, and B. This technical input output relationship has to be specified, for example a motor needs between 1,2 and 1,3 Volt as input from a power supply to perform its function. A modification in the function of module A may cause a change in the functioning of module B and the other way around. This brings us to our definition of the first type of technical interaction.

(1) Type 1: Two physical chunks do interact if: exchange of material, energy, or information between the two physical chunks is necessary to perform their functions.

The strength of type 1 interactions will depend on the range of possible values for the input and output. The more sensitive a function A is for small changes in the output of function B the stronger the interaction and the more integral the architecture. As a consequence, the stronger the interaction, the more coordination effort is needed during the design process.

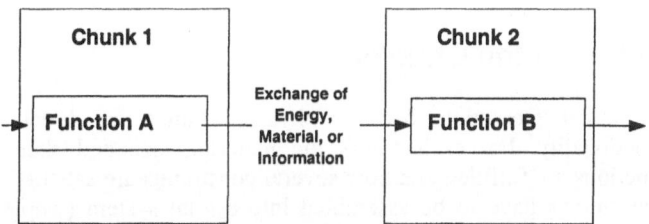

Fig. 6. Example of type 1 interaction

Next, the type of mapping between functions and physical chunks determines which physical chunks implements what functions. In case of a totally modular 1:1 mapping, each function is performed by one corresponding physical chunk. [29]. A functional requirement can be manipulated by a unique chunk, and a change in a chunk only effects its own function. Thus for the example in Fig. 5, the physical chunks 1 and 2 can be designed independently to perform their own unique functions A and B (if the type 1 interactions are determined). However, most products do not have a 1:1 mapping, and are more integrated because of technical or economic reasons. In these situations a change in a functional requirement may lead to changes in more than one physical chunk. Further, a change in one physical chunk may cause a change in another chunk because they both contribute to the same functional requirement. We define that when two physical chunks together fulfil the same function these chunks have a technical interaction, because a change in one chunk will require a change in the other chunk to fulfil the functional requirement. This leads to the following definition, which is also depicted in Fig. 7:

(2) Type 2: Two physical chunks do interact if: they together fulfil the same function (and this function is difficult to split up into independent sub-functions).

Of course, for totally modular designs no type 2 interactions exist for all pairs of modules. As a consequence for totally integrated architecture all (n) functions are fulfilled by all (m) chunks (n:m mapping), so all pairs of chunks then have a type two interaction for all functions. The more interactions of type 2, the more coordination effort will be needed during the design process.

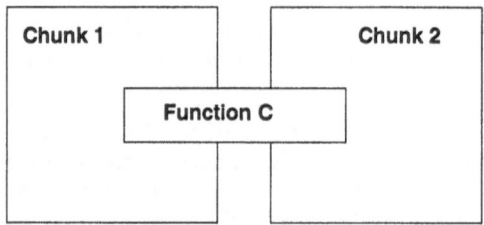

Fig. 7. Example Type 2 interaction by function C

5.3
Physical Integration of the Chunks

We elaborated the technical interactions between physical chunks that directly relate to desired functionality. For each chunk the material, size and shape determine how its functions are fulfilled and how several constraints are satisfied. However, all physical chunks have to be assembled into a total system (whole product) [2]. Hence the integration of the physical chunks can impose some additional technical interactions because of physical constraints or undesired effects. Technical interactions of type 3 appear if chunks physically have to be attached, or share the same limited amount of space and are in proximity of each other. Particular physical implementation of a function of a chunk therefore may interact with physical implementations of other chunks because of these pure physical constraints. For an existing product all chunks are physically arranged, linked, and share a restricted amount of space. Hence all kinds of technical interactions of type 3 can be documented, and we state the following, without claiming to be complete.

Space constraint
When chunks share a restricted amount of space, the size and shape of the chunks may cause technical interaction. The use of space of one chunk excludes the use of that space for another chunk. Therefore, a change in the shape or arrangement of one physical chunk may effect the physical properties of another physical chunk. Primarily, these interactions are pure physical and have nothing to do with

functionality. Of course, a necessary change of functionality of a physical chunk may influence its size or shape that in turn influences the available space for other chunks.

(3a) An interaction of type 3 may occur if desired changes in the size, shape, or position of a chunk will effect the size, shape, or position of the other chunk.
Side effects
Further, proximity of physical chunks may cause that undesired side effects of a physical chunk influences a neighbouring chunk. Undesired effects of a physical chunk may be unintended electromagnetic radiation, excess heat, vibration, etc. The closer physical chunks are arranged, the more sensitivity for side effects of each other. These effects cause technical interaction, because a change in a physical chunk may influence the physical properties of another physical chunk.

(3b) An interaction of type 3 occurs if undesired effects such as excess heat, vibration, etc. of one chunk, effect the functioning of another chunk.
Assembly
In the production process, physical chunks have to be assembled. If two physical chunks have to be assembled they share exactly the same constraints. A change in one module may necessarily cause a change in the other to make assembly possible and satisfy rules for Design for Assembly.

(3c) An interaction of type 3 occurs between two chunks if these chunks have to be assembled.
Type 3 interactions become stronger if less space is available, more side effects occur, and the more detailed the assembly constraints are. The stronger the interactions, the less modular a product architecture will be and the more time has to be spent on small changes in arrangement and detailed design of the physical chunks.

Now we have formulated the three types of technical interactions and its technical causes we are able to describe the managerial implications of documentation and visualization of these types in the next section.

5 Visualization and Managerial Implications

In order to formulate how technical interactions can be related to coordination effort in design processes, we will first describe how these interactions can be documented and visualized. Next, we argue how coordination effort between design tasks can be deducted. Finally, we will describe some managerial implications.

6.1
Documentation and Visualization

We propose that the technical interactions of an existing product can be documented in the following way. First, with consult of a team of designers the product has to be split up into clearly recognizable physical chunks. Similar to DSM, these chunks can be plotted into the rows and columns of a matrix. Second, for all physical chunks a designer has to be selected and asked for the functions that this chunk performs[30]. Third, for each pair of physical chunks we can document the technical interactions in a separate matrix for each type of interaction (as depicted in figure 8). This means that for example a necessary exchange of energy between chunk 1 and 2 is visualized in the element by documenting 'energy' in the element of the matrix (chunk 1, chunk 2). A type 2 interaction between chunk 1 and 2 can be visualized by documenting the common function that they have to perform together in the element (chunk 1, chunk 2) of the matrix. Finally a type 3 interaction between chunk 1 and 2 can be visualized by documenting 'assembly', 'the type of side-effect' 'or 'space' in the element (chunk 1, chunk 2) in cases the interactions occur. Together the three types of interactions are clearly recognizable, what is beneficial for a common and objective understanding. Further, for each documented interaction the technical cause can easily be identified.

Type 1 interact-ions	Chunk 1	Chunk2	Chunk 3
Chunk 1	X	• excha nge energy	--
Chunk 2	X	X	--
Chunk 3	X	X	X

Type 3 interac-tions	Chunk 1	Chunk2	Chunk 3
Chunk 1	X	• Assem-bly • heat	• Space
Chunk 2	X	X	--
Chunk 3	X	X	X

Type 2 interact-ions	Chunk 1	Chunk2	Chunk 3
Chunk 1	X	Func-tion C	
Chunk 2	X	X	--
Chunk 3	X	X	X

Fig. 8. Example of documentation of the three types of interaction.

6.2
Coordination Effort

Most likely the design of each documented physical chunk corresponds to the performance of sets of many interacting design tasks. For each chunk we will consider the corresponding set of design tasks together as a clearly identifiable aggregate tasks such as 'design chunk 1', 'design chunk 2', etc. This is plotted in Fig. 9 where aggregate tasks 1 and 2 relate to the design of chunk 1 and 2, and each exist of a set of smaller interacting subtasks. We assume that the hierarchical level of each aggregate design task corresponds to the hierarchical level of the physical chunk (that can be obtained by looking at the Bill of Material). In this way we constructed a set of tasks from which eventually hierarchical differences can be recognized easily.

Now, if a pair of physical chunks (i.e. chunk 1 and 2) do not have technical interactions, the two corresponding aggregate design tasks can be performed independently. It logically follows that if the two modules do have a technical interaction this causes a need to exchange information between the aggregate design tasks, which have to be coordinated. So documentation of the technical interactions between all pairs of modules gives the technical causes for information exchange between all aggregate design tasks. Note that we cannot discover the direction (serial or reciprocal) of necessary exchange of information between aggregate design tasks by looking at the product alone and therefore cannot perform traditional (mathematical) DSM analysis. However, for each type of interaction we have an indication about the coordination effort that is needed. The documented technical interactions can be presented to groups of designers of a design project, and they can asked for their experience with necessary exchange of information between the aggregate design tasks. The effort to coordinate a technical interaction and the amount of iteration that occurred between the aggregate can be discussed and formulated for each documented type of interaction. This knowledge about the technical interactions can be applied to improve and accelerate future projects.

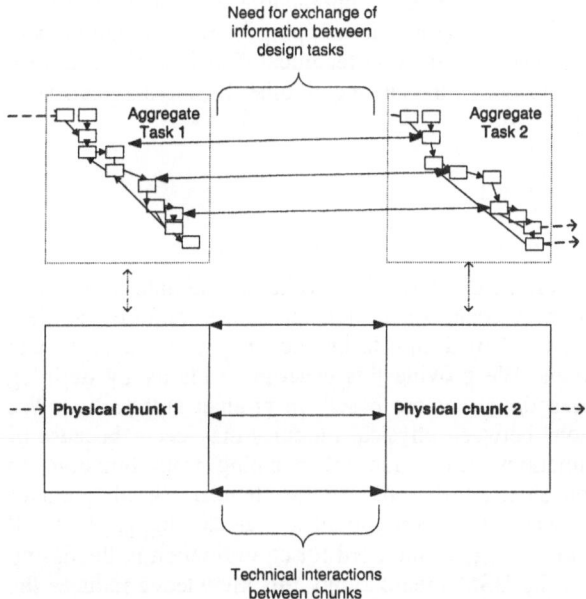

Fig. 9. Linking technical interactions between product modules to need for coordination between design tasks.

6.3
Improvement

For a new design process that is based on the architecture of an existing product it is likely that most types of interactions between physical chunks will return. For instance the power supply and the motor still have to exchange energy. However, the exact specification of this interaction may be subject to change. For similar types of interactions the knowledge of the previous design process can be applied to improve coordination. For example improved scheduling of design tasks, improved assignment of design tasks to design groups, Set-based concurrent engineering, simulation studies to predict side-effects, meetings, change protocols, standard design rules etc.

Another way to improve future design processes is to manipulate technical interactions in order to reduce the need for coordination. A type 3(a) interaction can be reduced if alternative physical chunks or components are selected with similar functionality, but with a smaller size such that relatively more space is available. Similarly, by specifying a broader range of possible functional inputs and outputs of chunks can strongly reduce the strength of the interactions of type 1. This over specification may reduce the coordination effort that is needed for type 1 interactions. In this way a physical chunk may become less dependent on possible changes in specifications of exchange of energy, or material from another chunk. Of course possible extra unit costs to buy or produce these components should be balanced against reduced need for coordination and possibilities for standardization. Further, type 2 (and type 1) interactions can be adjusted by changing the function structure or changing the mapping between functions and physical chunks. However, adjustment of these technical decisions needs a very careful consideration of the influence of the change on other business aspects such as production and marketing.

7 Conclusions

In this paper we argue that current DSM studies are hardly suitable to improve future design processes. Improvement of design processes requires a clear indication of the hierarchical level of design tasks and insight in the cause and nature of technical interactions. We provide this descriptive clarity by defining three types of technical interactions that are based on product architecture. We state that technical interactions between physical modules may occur because of (1) interactions between functions, (2) a non 1:1 mapping from functions to physical chunks, and (3) physical attachments of the chunks. Visualization of these types of interactions between all pairs of modules of an existing product will provide clear insight in and knowledge of the need for co-ordination in the design process. In combination with the DSM methodology this knowledge induces the improvement of *future* design processes. Systematic learning about how technical interactions have been coordinated creates opportunities to improve coordination of design processes based on the same types of technical interaction. Further, understanding of the causes of technical interactions gives structural insight in options to significantly change a product architecture in order to manipulate the most severe technical interactions and reduce the need for coordination.

Recently, the practicality of using this method and typology is demonstrated for a design process of an electric shaver. The visualization of technical interactions gave significant insight in the complexity of this design process, and clearly gave the users (and us) more insight in the managerial implications of technical interactions.

8 References

[1]   Wheelwright, S., & Clark,K.B. (1992) *Revolutionizing Product Development* (1 ed.). New-York: The Free Press

[2]   Pahl, G., & Beitz,W. (1996) *Engineering Design: A Systematic Approach* (2 ed.). London: Springer

[3]   Cooper, R.G. (1993) *Winning at New products: Accelerating the Process from Idea to Launch* (1 ed.). Addison-Wesley

[4]   Ettlie, J.E. *Product-process Development Integration in Manufacturing.* In Management Science, Vol. 41, N 7, 1995 . pp. 1224-1237

[5]   Gell-Mann, M. *What is Complexity* . In Complexity, Vol. 1, N 1, 1995

[6]   Anderson, P. *Complexity Theory and Organization Science.* In Organization Science, Vol. 10, N 3, 1999, pp. 216-232

[7]   Simon, H.A. (1981) *The Science of the Artificial* Cambridge Massachusetts: MIT Press

[8]   Orton, J.D., & Weick,K.E. *Loosely Coupled Systems:A Reconceptualization.* In Academy of Management Review, Vol. 15, N 2, 1990. pp. 203-223

[9]   Thompson, J.D. (1967) *Organizations in Action* New-York: McGraw Hill

[10] Galbraith, J.R. (1973) *Designing Complex Organizations* Reading: Addison-Wesley Publishing

[11] Von Hippel, E. *Task Partitioning: An Innovation Process Variable.* In Research Policy, Vol. 19, 1990, pp. 407-418

[12] Steward, D.V. *The design structure system: a method for managing the design of complex systems.* In IEEE Transactions on Engineering Management, Vol. 28, N 3, 1981. pp. 71-74

[13] Smith, R.P., & Eppinger, S.D. *Identifying Controlling Features of engineering Design Iteration.* In Management Science, Vol. 43, N 3, 1997 . pp. 276-293

[14] Smith, R.P., & Eppinger, S.D. *A Predictive Model of Sequential Iteration in Engineering Design.* In Management Science, Vol. 43, N 8, 1997 . pp. 1104-1120

[15] Smith, R.P., & Eppinger, S.D. *Deciding between Sequential and Concurrent Tasks in Engineering Design.* In Concurrent Engineering: Research and Applications, Vol. 6, N 1, 1998 . pp. 15-25

[16] Smith, R.P., & Morrow, J.A. *Product Development Process Modeling.* In Design Studies, Vol. 20, N 3, 1999, pp. 237-261

[17] McCord, K.R., & Eppinger,S.D. *Managing the Integration Problem in Concurrent Engineering.* In Working Paper MIT Sloan school of Management, Vol. 3594, 1993 pp. 1-48

[18] Wood, R.E. *Task Complexity: Definition of the Construct.* In Organizational Behavior and Human Decision Processes, Vol. 37, 1986 . pp. 60-82

[19] Gulati, R.K., & Eppinger, S.D. *The Coupling of Product Architecture and Organizational Structure Decisions* (1996). Anonymous. Sloan Working Paper. Cambridge, MA: MIT. 3906

[20] Staudenmayer, N. *Webs of interdependency: strategies for managing multiple interdependencies in new product development.* In Proceedings of the EIASM 6 th international product development conference 99 A.D. Cambridge U.K.: University of Cambridge, Institute for Manufacturing. 989-1007

[21] Eisenhardt, K.M., & Tabrizi, B.N. *Accelerating Adaptive Processes: Product Innovation in the Global Computer Industry.* In Administrative Science Quarterly, Vol. 40, 1995 . pp. 84-110

[22] Ulrich, K. *The role of product architecture in the manufacturing firm.* In Research Policy, Vol. 24, 1995 pp. 419-440

[23] Erixon,G. (1998). *Modular Function Deployment- A Method for Product Modularisation.* Thesis

[24] Ulrich,K. (1991). *Fundamentals of product modularity.* Issues in Design Manufacturing Integration ASME, 39

[25] Sanchez, R. *Product and process architectures in the management of knowledge resources* (1999). Anonymous. Working Paper. Lausanne Switzerland: International Institute for Management Development. IMD 99-6,

[26] Henderson, R.M., & Clark,K.B. *Architectual Innovation: The Reconfiguration of Existing Product Technologies and the Failure of Established Firms.* In Administrative Science Quarterly, Vol. 35, 1990 pp. 9-30

[27] Sanderson, S., & Uzumeri, M. *Managing product families: The case of the Sony Walkman.* In Research Policy, Vol. 24, 1995 pp. 761-782

[28] Whitney, D.E. *Why Mechanical Design cannot be like VLSI Design.* In http://web.mit.edu/ctpid/www/whitney/morepapers/design.pdf, Vol. 1996

[29] Albano, L.D., & Suh, N.P. *Axiomatic Approach to Structural Design.* In Research in Engineering Design, Vol. 4, 1992 pp. 171-183

[30] Pimmler, T.U.,& Eppinger, S.D. (1994). *Integration Analysis of Product Decompositions.* In ASME Design Theory and Methodology Conference 0 AD Minneapolis: 343-351

Part II

Development of Product Portfolios and Module Systems

Record from 2nd Group Work

Antti Pulkkinen

Tampere University of Technology
P.O.Box 589
33101 Tampere, Finland
e-mail: pulkkine@ruuvi.me.tut.fi

1 Introduction

The group work section followed the presentations of the papers by Smith & Duffy, Lange & Åslund, and McKay et al. Thus, the discussion was very much related to issues of knowledge capturing and re-use, quality, and data structuring. Some of the highlights from the discussion have been captured in the following pages.

2 Records from Discussion on Development

The records consist of four sections. First some opinions on product structuring purpose and goals are described. Then some quality issues are depicted. Third section is a transcription of a debate on knowledge applicability and quality issues. Finally, some remarks on terms and definitions are included.

2.1
Structuring purpose and goal

The purpose of product structure development can be seen as a way of turning data into information and further to knowledge. Several sub-activities or synonyms of data structuring activity were recognised. These were information systemization, classification and formalisation. It was suggested that the data resolution level (i.e. the degree of details) is related to process that uses or produces the data.

However, data structuring is not an adequate reason for a resource and time consuming structuring project. The development project has to be justified with effects it creates in activities and properties it causes for products. The main effects discussed were knowledge re-use, increased organisational efficiency in carrying out activities in a more rational structure and design co-ordination possibilities.

The goals of product structure development projects vary. Some participants emphasized that the goal is a product platform while others stressed product architecture or re-usable knowledge. At least a one participant argued that the eventual objective is not any of these, but rather the benefits in running business. Furthermore, he suggested product structuring is not a design office issue, but a business management issue. However, product platforms and architectures, configuration models, and product stucturing in general are means for achiving the business objectives. Developing the means requires a number of distinct viewpoints.

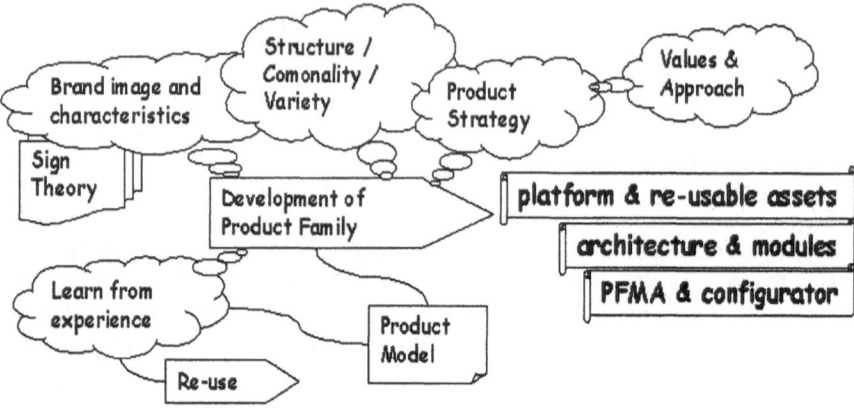

Fig.1. Different viewpoints to the development of product families

In Fig. 1 the idea of product development goals as a tool for business is sketched. All the participants had different opinions on useful methods and theories to be used in product structuring projects (represented by thought bubbles) depending on their experiences. Because their goals were different, some saw the others' goals as means for the project. For example re-use is at the same time a way to learn from experience and a goal for taking advantage of structuring (see lower right corner of the Fig. 1).

It is necessary to regocnize that a product structuring activity leads to structure that can (and should) be re-used in other activities. It was stated that at least two different kind of product development projects will exist. There are product developing activities, which develop the structure and re-usable assets, and activities, which harvest the structure and assets. So, the development projects lead to another projects depending on the evolving business objectives.

2.2
Quality

The presentation on "Capturing Quality Perceptions..." stimulated some participants to raise questions on applicability of the approach. It seems the number of quality perceptions depends on the number of persons interacting with the product. Taking this into account, a practical dilemma is the adequate guantity of perceptions needed to investigate for creating a general and well defined model of product quality. In extreme case, does all the perceptions have to be captured independently?

It was also doubted if applying sign theory is possible in business-to-business sales and related products? Also the meaning of gestalt in product structures was unclear. Does it mean the inherent perceptions on product structure (i.e. the default structure in each persons' individual "product model")?

2.3
Knowledge Applicability and Quality

One of the most notable purposes for product structuring is the acquisition of product structure related knowledge and possibility to re-use the knowledge. However, knowledge re-use tends to raise some questions like how to recognize the knowledge that is worth of acquisition (i.e. the knowledge that is generally valid)?

Careful attention has to be paid on the quality issues in a knowledge acquisition task. In a discussion, several questions were addressed on the quality of re-usable solutions. What are the evaluation aspects when capturing knowledge and deciding on re-usable solutions? What is a good solution and what is a poor one? How to identify and separate the good solutions from bad ones? Is there a measure for knowledge correctness?

One group recognised that the knowledge may not be applicable when changes occur. The design problem may be different from the initial one (i.e. the problem was different when knowledge was captured) and the number of possible solutions increase with technical evolution. How to measure the adaptability between the existing knowledge and the problem at hand? What is the knowledge context and how to represent it? Should the rationale behind re-using a solution be visible or hidden from the user?

2.4
Re-use Contents

One of the inherent properties and purposes of structuring activity is design re-use. It has been stated (e.g. later in this proceedings) that activities and knowledge are related to product structures. However, the relation between re-use and these issues is not clear. Are the activities going to be re-used as well as structural elements (relations, modules, components, parts, features, etc.) and knowledge related to structures? Or is it only the results of activities (i.e. structure and knowledge elements) that are re-used? What about re-using activity structures in planning?

These questions have theoretical values, but they are important also in practise. When developing product models a developer has to make decisions on re-usable assets and to know what kind of relations are needed to establish. Therefore, it is important to recognize the purpose, viewpoint, and the context of the model.

A practical opinion, which was given, suggested to learn from the experiences in software engineering. For example methods for change management have been created in software design. It was suggested these methods could be re-used further in product structuring for re-use. Practises from object-oriented analysis and object-oriented design were touched in discussion. Another interesting issue might be to use design patterns (earlierly developed by architecturel and software engineering).

The relation of activities and re-usable assets was discussed. Especially the benefits of modelling activities where questioned. Is an activity model a descriptive or prescriptive model? How these alternatives affect on making relations with re-usable assets? It was also questioned if activity modelling enhances design co-ordination? How to decide between re-use and innovation actions?

3 Terminology

It was stated that each person has his/her own experiences on product structuring projects and perceptions on the projects' purposes. These projects are also related to different contexts of technology and business. The outcome is a tangled web of definitions and terminology. E.g. in the paper "A framework for evaluating commonality" seven propositions to the term product platform has been given.

It was stated an unified terminology is needed for different product structuring issues. The terms listed in workshop, can be grouped into four classes.
1) Product structure system.
2) Entities of the system.
3) Attributes of the entities and the system.
4) Relations between the system elements.

The recognized first class terms are Product Platform, Framework, Architecture, Portfolio, Program, and Structure. Second class terms used in dicussions were Module, Chunk, and Reusable asset. The terms belonging to third class depicted similarities and uniqueness. The used terms were Variance, Diversity, Familiarity, and Commonality. The fourth (relations) class terms Interface and Interaction were also used.

Product Structuring for Design Re-use

J. S. Smith, A. H. B. Duffy

CAD Centre
University of Strathclyde
75 Montrose Street
Glasgow, UK
GI, IXJ
e-mail: jo@cad.strath.ac.uk, alex@cad.strah. ac.uk

Abstract. The contemporary view of design re-use reflects the utilisation of any knowledge gained from a design activity and not just of past designs of artefacts. Structuring knowledge from the design process is an essential part of design re-use. Without structure, utilisation of knowledge re-use would prove more difficult. The product structuring principles of decomposition, configuration and rationalisation go part way to meet the requirements of 'design re-use'. The process of *'designing for re-use'*, however, continues to cause difficulties for the design research community. The effective capture of often abstract knowledge from an *'evolving design model'* for subsequent re-use requires further research. Structuring principles play a large role in this, in that the applicability of any repository of knowledge is limited without adequate structure. Despite this, a mapping process of product structuring to design re-use identifies that *'design for re-use'* lacks support from product structuring principles. The potential for enhancing a number of structuring principles to facilitate *'design for re-use'* is highlighted with the specific example of modularisation.

1 Introduction

The goal of this paper is to inform the design research and product structuring community of both the over-riding theories behind design re-use and the role of product structuring within this field. One of the central aims in meeting this goal is an exploration of how current product structuring research and a proposed future direction serves to meet the requirements of a re-use approach to engineering design. An existing 'design re-use process model' [1] is used as a base over which to map current research. This mapping process facilitates a deeper understanding of the components of design re-use, where distinct structuring approaches reside, and their interactions within an overall design re-use approach.

The following section gives a brief introduction to design re-use, identifying it's potential to provide a powerful approach for harnessing experiential design knowledge for use in future design scenario's. The potential benefits of re-use as an approach applicable before, during and after design are posited. Section 3 focuses on the current role of product structuring with respect to re-use. Prevalent approaches are mapped to a 'design reuse process model', to highlight current support for it's processes and components. The problems and potential of product structuring in providing support for design re-use are considered. Section 4 expands on this potential and examines the ability of product structuring to sustain '*design for re-use*', the process shown as most deficient with respect to formalised support. Finally Section 5 concludes the paper by summarising the main findings and posing a number of questions raised by the need to effectively support 'design for re-use'.

2 Design Re-use

Re-use has become a significant topic within the design research community as reflected in the 1998 Design Re-use, Engineering Design Conference. Most notably, this conference highlighted the differences in perceptions as to the nature, role and applicability of re-use in engineering design.

In the foreword Sivaloganathan [2]claimed that 'based on the hypothesis that increased use of subsystems and parts that have proven to be successful in past designs [re-use] is a powerful way to enhance the ability to produce designs that are easy and cost effective to manufacture with guaranteed performance'. Culley [3] saw the 'use of standard parts in design as the classic case of design re-use in that designer's know that they are getting a well proven product with a wealth of development and 'field' experience behind it.' Both these statements emphasise the re-use of specific parts or subsystems, Culley however also [3] noted that '[standard components] are being developed.... to enable both the re-use of the part and the experience associated with that part'. Finger [4] further extends this notion by stating that 'designers may re-use a prior design in it's entirety by selecting it from a catalogue, may re-use an existing shape for a different function, or may re-use a feature from another design'. Where previously re-use has tended to mean the direct utilisation of past designs, the concept of reusing experience and previously acquired concepts as well as objects in a new situation, is introduced. Re-use can now be considered 'to reflect the utilisation of any knowledge gained from a design activity and not just past designs of artefacts' [5].

Gao, Zeid, Bardasz [6] estimate that 90% of industrial design activity is based on variant design and in such a redesign case an estimated 70% of the information is re-used from previous solutions' [7]. Hence, when a new design problem arises, it is frequently solved by modifying existing designs rather than creating a completely new design. Characteristically designers do not 'reinvent the wheel' every time a new design instance calls for one, their natural response is to glean from past experience and re-use previously acquired knowledge.

We see the concept of re-using as inherent within the natural process of design. However, the origins of formal design re-use practices can be found in the realms of software engineering where designers, 'faced with increasing complexity and time-to-market pressures, began to consider re-use as a realistic solution to their problems'[8]. Engineering companies, also faced with increasing competition from global markets and greater consumer demands in terms of cost, time and quality, are consistently find relying on the designers natural inclination to re-use as a means to meet such demands, insufficient and complacent.

2.1
Why Re-use?

Sreeram and Chawdry [9] claim that, due to the potential for use of well tested and optimised concepts and objects, engineering design re-use can be utilised to fulfil the requirements of decreasing design times, increasing design quality, improvements in the predictability of designs and reduced costs. Time, cost, quality and performance were amongst the main benefits analysed in a study by Duffy and Ferns [10]. The study concluded that the potential benefits to an industrial company, of applying an overall re-use approach, far exceeded the benefits they currently received from their ad-hoc application of re-use which relied on designers' natural inclination to re-use. Figures 1 and 2 provide a summary of their findings.

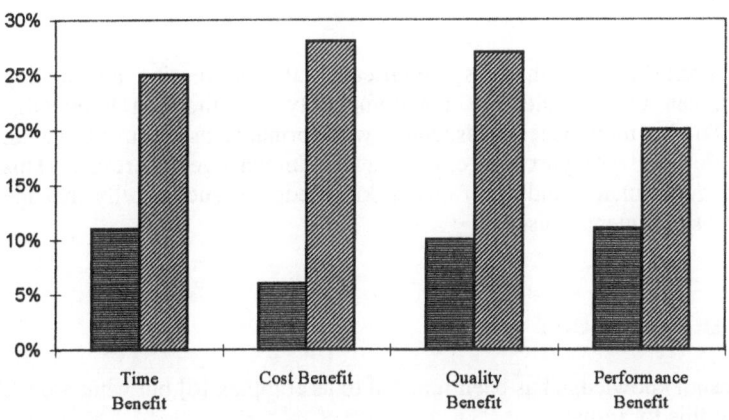

Fig.1. Current and Foreseen Overall Metric Benefits [10]

The first column, in each category (time, cost, quality and performance), in Fig. 1, shows the benefits received from current ad-hoc re-use practices which provide the most benefit to time and performance whilst the costs benefits of re-use are almost half at only 6%. The second column, in each category in Fig. 1, indicates the foreseen benefits of a formal approach to design re-use which is expected to provide time, cost, quality and performance benefits in the region of 20% to 28%. This can be translated into an improvement in terms of cost benefits of up to 367% over that received from current practice (as shown in Fig. 2). The study verifies that those re-use practices naturally inherent in the process of design do provide benefits. More importantly, the results substantiate the need for formalised approaches, methodologies and systems, to achieve the considerable potential benefits available as a result of re-use in design.

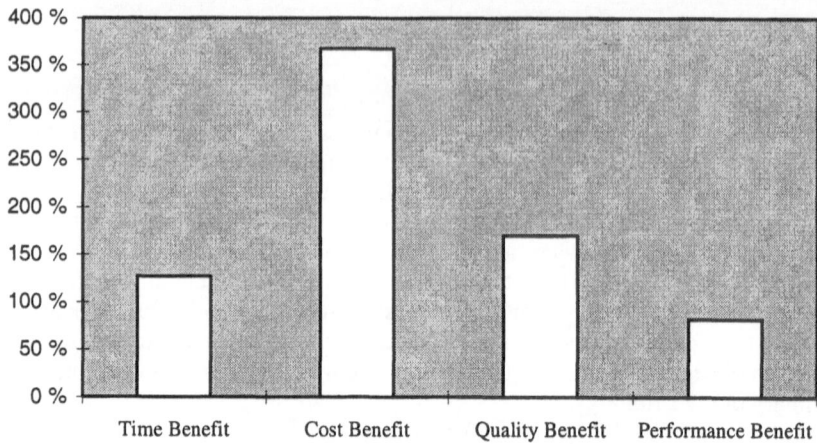

Fig. 2. Foreseen Metric Improvements [10]

We have established that there is 'significant value in reusing an existing design' which can be attributed to 'the complexity and the rich knowledge involved'[6]. To obtain the benefits associated with formal re-use requires that we optimise the rich and complex sources of design knowledge for re-use. This involves an in-depth understanding of design knowledge to successfully manage it's capture and subsequent re-use.

2.2
What can be Re-used?

Previously, design knowledge has been referred to as complex [6] but what are the implications of this for re-use?

The concept of complexity is one that is ill defined. Definitions of complexity are highly dependent on the context in which it is discussed. We can begin to establish the difficulties of supporting design knowledge for re-use where re-use relies on the ability to explicitly define a formal representation of complex design knowledge. Forming a deep understanding of the nature and growth of knowledge is essential for implementing better systems and more effective re-use practices.

The simplest product can have a large quantity of information identified with it, for instance, a simple pen with only 4 component parts has associated knowledge about geometry, polymers, ergonomics, safety standards, fluid flow, manufacturing processes, suppliers, users and markets, etc. The problems of modelling such knowledge for re-use are amplified by the differences in the way this information is presented in such terms as type, content, scope, completeness, abstraction level, refinement and media. Thus, the management of a product's complexity is a key feature in a successful re-use approach.

Design knowledge is produced continually throughout the design process, though 'typically the only formal documentation is the final set of drawings' [4].

Generally these contain knowledge of both the product and manufacturing processes, they are based on a low level of abstraction e.g geometry, tolerance, surface finish, manufacturing requirements. Such formal documentation leaves little or no scope for representing knowledge related to the rationale, history, or product knowledge relating to concept principles and dynamic process knowledge learnt through experience. Hence, the designers capabilities of identifying 'good' design elements, appropriate for re-use, is minimised by a lack of understanding of how and why the final design solution was attained. As illustrated by the light grey arrows in Fig. 3, such low level knowledge forces designers to think of design specifics which have limited applicability to the earlier synthesis stages of design and limits re-use principally to support of the detailed design stage. Eighty percent of the life phase costs of a product have already been determined by the end of the conceptual design stages [11].

High level knowledge about the function, behaviour and rationale of a product are essential to the re-use approach in that it can facilitate both the assessment of a product's complexity and effective management of this. High level knowledge can consolidate low level information about geometry and physical characteristics by providing the user with a deeper understanding of how, why and what elements combine to produce a design. This 'deeper' understanding increases the potential for identification and abstraction of 'good' design or knowledge elements with greater re-use capabilities.

Much of this high level knowledge is produced earlier in the design and provides the foundations on which the specifics of a design are formed. Capturing such high level knowledge can facilitate a re-use approach applicable far earlier in

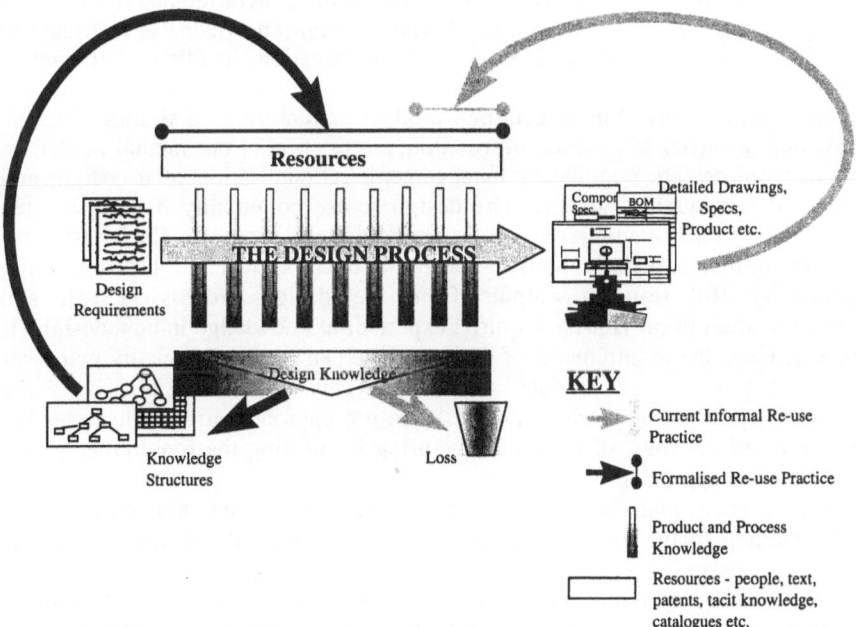

Fig. 3. The Impact of Re-use Practice on the Design Process

the design process, highlighted by the black arrows in Fig. 3. Currently however, the motivation to document the design and decision processes is confined to individual designers and their notebooks. Accordingly, this part of the process is generally not formally supported and as a consequence the knowledge is not made explicit and is displaced to the memory of the individual. Fig. 3 illustrates schematically the impact of current re-use, where knowledge capture is limited to low levels, forcing designers to work in specifics with limited applicability in earlier design stages (grey arrows and lines in Fig. 3). A formalised re-use approach (shown by the black arrows and lines in Fig. 3), where knowledge capture throughout an evolving design process is promoted, can extend the applicability of knowledge for re-use throughout the design process.

To benefit fully from the re-use approach, as defined in this paper, we must understand design knowledge, appreciate it's nature and how it develops and grows through the design process. Such understanding of design knowledge is a prerequisite to effective re-use. Only then can we provide the structures to manage it's growth, contain it's complexity, and harness both the high and low level knowledge necessary to fully define a design and design process for re-use.

3   Product Structuring in Design Re-use

Product Structuring concerns the activity whereby the structure characteristics of a design or artefact are defined. Where structure is defined as 'the elements of a system identified by their type and relations between these elements' [12]. Not purely limited to physical parts and components, the structuring activity can also be utilised at a far more abstract level where design knowledge is restricted to high level knowledge describing energy transformation, functions, behaviours, etc.

The motivations for executing product structuring activities include providing: a carrier of product information; a reflection of our mental model; an output from design; complexity management; rationalisation of a design; and improved communication [13]. The design re-use community have also cited (section 2.0 to 2.3) a number of these as important requirements. Such design re-use requirements include housing design related knowledge [1], managing complexity [6], re-using outputs from the design process [1, 14], and communication of previously acquired experiential knowledge in new design [1, 10, 14]. Thus, the requirements of design re-use can be seen to readily map over the driving forces of product structuring. This correlation highlights the applicability of product structuring to the re-use approach. Fig. 4 illustrates the need for and the roles of product structuring in meeting the requirements of a design re-use approach.

It is argued that successful product structuring relies not only on an understanding of the theories and methodologies behind it but also how it fits within the design process, the product development strategy and overall company strategy (see Fig. 4). For a strategy geared towards continued enhancement and re-use of a company's knowledge resources, the role of product structuring is that of supporting, maintaining and promoting this resource for re-use.

Figure 4 illustrates the roles of product structuring (P.S) and re-use within a design organisation. It shows how the overall organisational and product development strategies constrain the design team throughout any design project. The diagram depicts both structured and unstructured re-use and highlights the need for tools, techniques and methodologies to rationalise from past design cases and structure current design knowledge to facilitate a structured approach to re-use and learning. Both the structuring and re-use principles must also develop to satisfy organisational goals whilst meeting the knowledge requirements of the design team in both current and future design projects.

The design team are 'pulled' by the overall company and product development strategy and in turn 'push' the individual design project by carrying out a number of known processes. The design team are also subject to a number of influences including their own knowledge and experience. Thus, a design team draws from past design experience and re-uses experiental knowledge to further the current design project within the overall design requirements.

Product Structuring is essential to a re-use approach in that design generally produces a vast amount of knowledge, information and data (K.I.D), which without structure remains a mass of vague elements whose significance and relations are difficult to define and understand. With structure however, this K.I.D can aid designers in finding specific knowledge elements useful in a new design i.e. Design by Re-use. Such K.I.D from past designs can be structured using a number of different principles from viewpoints to families and inter-linked with a series of networks.

If product structuring is then employed during current or new design processes for products, parts, assemblies and families the K.I.D generated during this evolving process can be grouped and stored in a formal manner, and appropriately structured to promote it's use in future design, i.e. Design for Re-use. As shown in the diagram we would thus have a bi-directional flow between past and current design which would result in structuring principles to support learning from design knowledge and it's subsequent re-use. Thus, Fig. 4 illustrates that product structuring is indeed an essential element in the re-use approach as product structuring can facilitate the organisation of K.I.D. However, current re-use in design organisations (as depicted by the solid black arrows) is predominantly unstructured [10], with a deficiency of tools, techniques and methodologies to support the overriding product structuring theories. Hence an increased understanding of the processes of product structuring is required to enable this K.I.D to be structured in design organisations for re-use. For instance, Duffy and Leglers work [14] addresses this need by proposing a methodology to structure (rationalise) past designs. This rationalisation provides a basis upon which to efficiently retrieve specific cases for re-use and presents a means upon which to generalise, enhance and re-use past experiential knowledge.

The following section outlines a design re-use process model over which current product structuring principles will be mapped in paragraph 3.2. This mapping process identifies the applicability of a number of contemporary structuring principles and highlights gaps in current support for re-use.

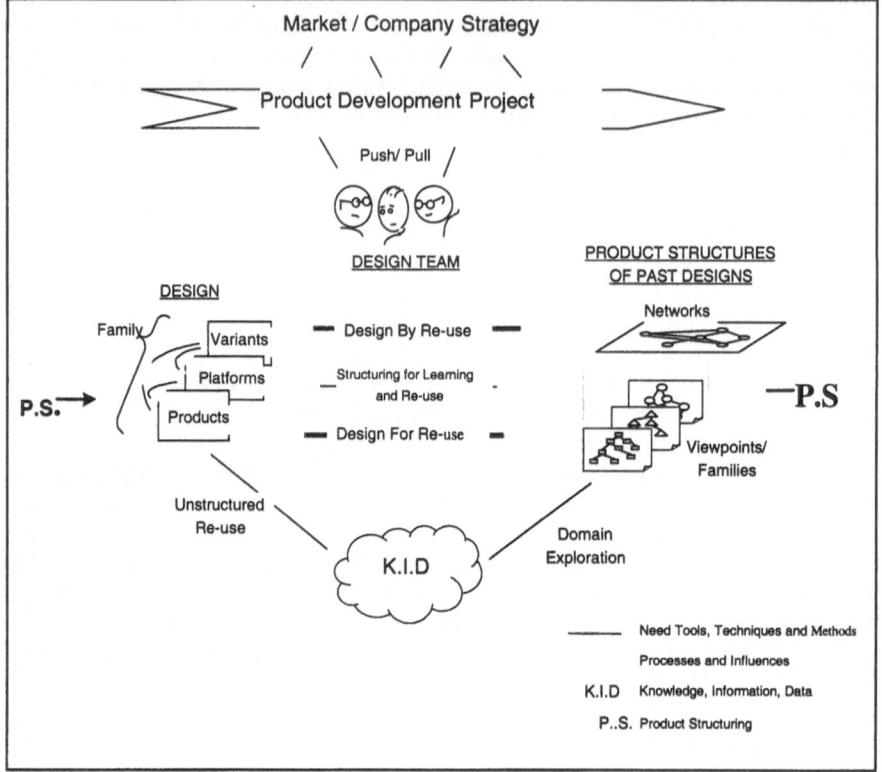

Fig. 4. Organisational Product Structuring and Re-use

3.1
Design Re-use Process Model

The model is made up of six knowledge components and three processes, brief descriptions of each are given below. A more detailed description of the generation and significance of the 'design re-use model' and the components and processes encompassed within it, can be found elsewhere [1, 15].

The re-use processes are[1]:

Design by Re-use – the re-use of previously acquired concepts in a new design situation. Design by re-use can only occur if reusable resources are available through, for example, 'domain exploration' and 'design for re-use'.

Domain Exploration – The examination of a design's domain knowledge from which reusable fragments can be identified, rationalised (structured), extracted and stored for subsequent use in developing new designs.

Design for Re-use – This process is carried out during design itself. It concerns the identification and extraction of possible reusable knowledge fragments of a current design and the enhancement of their knowledge content, including developed design alternatives, modifications and associated reasoning behind design decisions.

The knowledge components are [1]:

Design Requirements - a statement of design need.

Domain Knowledge - knowledge pertaining to a design domain, e.g existing product information, past design alternatives, potential solution alternatives.

Re-use Library - An organised collection and store of knowledge.

Domain Model - A representation of the designer's conceptualisation of the current design problem domain.

Evolved Design Model - A statement of an evolved design, which may be at any level of abstraction, of an incomplete design or a final completed design.

Completed Design Model - A completed statement of the finished design solution, which meets all the design requirements.

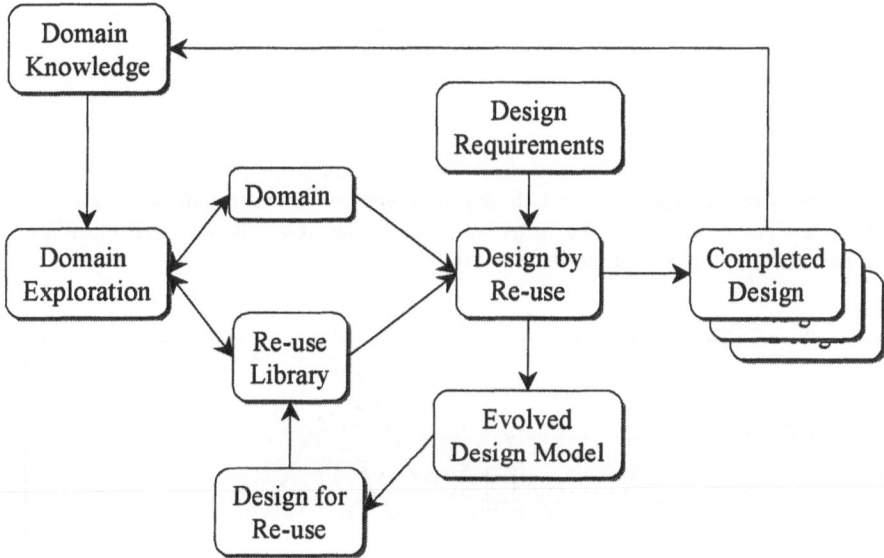

Fig.5. Design Re-use Model [1]

3.2
Structuring to Support Design Re-use

The basis of structuring is to define the elements and relations of a product or artefact with respect to a chosen viewpoint. A number of theories dominate structuring methodology including Andreasen's theory of domains [12, 16] and the model based theory subscribed to by Erens [17]. Andreasen's structuring principle centres around the synthesis process as a progression from transformation structure (energy, material etc.), through functions (required effects) and organs (function carriers) to the definition of parts structures. The basis of structuring principles, here, involve the definition of structures within each domain and understanding the interactions between domains (as shown in Fig. 6).

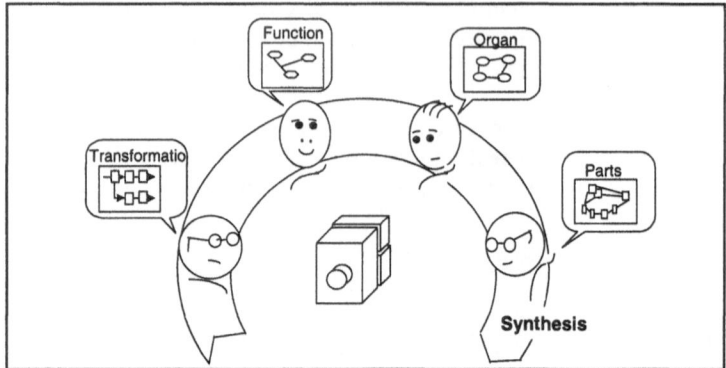

Fig. 6. The Domain Theory

Erens' product model based theory involves the transformation of a design from a functional model to a technology model towards the definition and construction of a physical model (see Fig. 7).

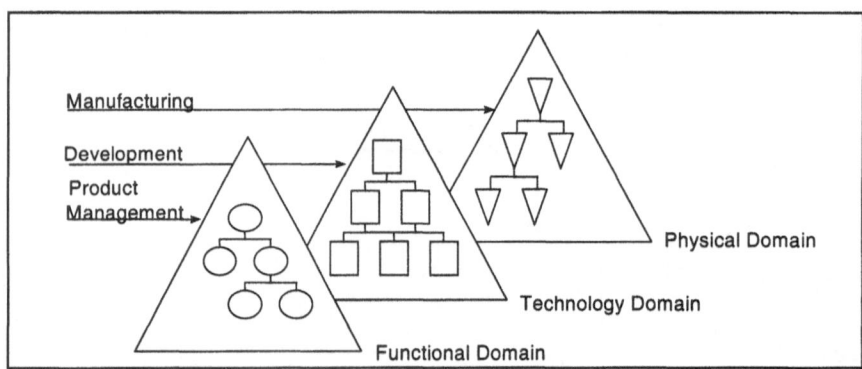

Fig. 7. Product Models

Both theories structure products in various domains but it is the scope, role and definition of each domain which differ. In general the structuring principles utilised within each theory are the same and it is merely the application and entry point into the process which differs. Structuring tends to fall into the categories of decomposition [18, 19], configuration [16, 18, 20] (occasionally referred to as composition [21]), rationalisation [22, 23].

An artefact, product, primary effect, primary function, or technology can be decomposed into it's constituent elements. This decreases associated design complexity because the integral complexity of each individual element is lower than that of the whole[18]. Thus, teams of designers can work in parallel and reduce product development time [18]. Decomposition is often related with the functional domain [16, 23] or effects domain [16] where the main function is decomposed into sub functions [21]. Liedholm [19] states that the decomposition activity is utilised to clarify what the product should do and establish the functions of the product.

The decomposition activity primarily supports 'design by re-use' in that it allows designers to breakdown the low level design requirements into more manageable, less complex constituents. Applied to the re-use library, the decomposition activity can breakdown past design solutions into their high level solution concepts providing the potential of utilising both experiential knowledge and design specification goals. Thus, designers can decompose current requirements with a view to map between current and past designs to enhance 'design by reuse' capabilities.

The elements of a product and the way in which they are built together determine the overall behaviour or function of that product [16]. Where a design need is decomposed into low complexity elements, often at an abstracted level, we must proceed to allocate possible solution concepts to each element and build these solutions back up to meet this need. This 'building up' activity is commonly referred to as configuration. Thus, configuration creates an arrangement, from a given set of elements, by defining the relationships between selected elements that satisfy the requirements and constraints of a design [20]. The process of configuration design involves the creation or identification of relations between the elements to ensure that the subsystem realises it's function and contributes to the overall purposeful function, in the right manner [18].

The activity of configuration can be seen to support 'design by re-use' as it re-uses previously defined elements to meet a current design need. Configuration as with decomposition can only support 'design by reuse' when rationalised (structured) sources of past design knowledge are available.

Rationalisation in product structuring involves the systematic organisation of product structure related knowledge to form a rational conception of a model which is free from radical or specific quantities. Rationalisation can take a number of forms including the definition of product architectures [22, 23], platforms [24] etc. Generic product architectures are defined to form a 'stable structure and provide a consistent environment for new component development [23]. Such product architectures arise from rationalisation over a number of products, are a more stable model of design than the physical models and can be re-used to create new versions of the product [23]. Where physical models consist of components which are liable to change, generic product architectures facilitate mapping of a more consistent functional arrangement to physical components and the interactions between these [22]. Rationalisation can also occur when components within a number of products are redefined to produce a platform of products. This platform forms a 'common structure from which a stream of derivative products can be efficiently developed and produced' [24]. Platforms form a re-usable 'foundation of product elements, technologies, knowledge as means of supporting product variety and increasing re-use of engineering knowledge' [24]

Rationalisation in structuring promotes the process of *'domain exploration'* by exploring completed design models and their associated domain knowledge. Such exploration results in a deeper understanding of the elements and relationships which combine to facilitate effective design in the domain. Successful rationalisation of design knowledge can promote the *'re-use library'*, of parts, concepts and knowledge and a generic product architecture model which can subsequently be utilised through the *'design by re-use'* process.

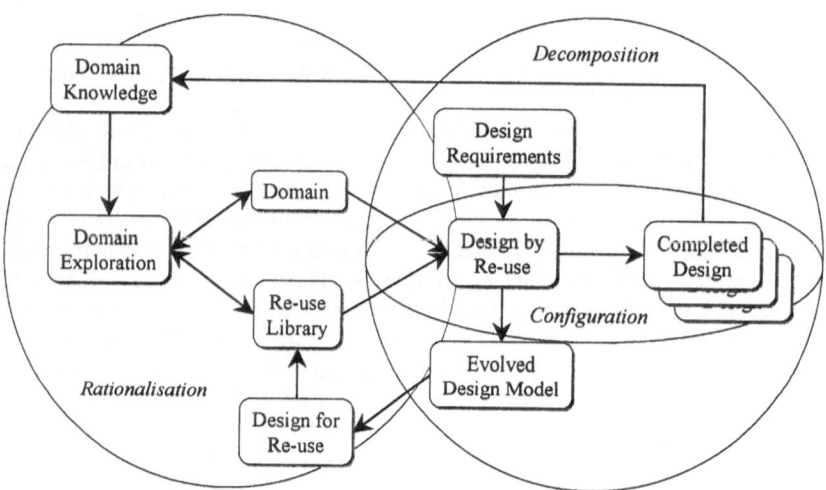

Fig. 8. Schematic View of Product Structuring Mapping

3.3
Current Structuring Potential and Problems

It can be seen from Fig. 8 that current product structuring principles predominantly support *'design by re-use'* with limited provision for *'domain exploration'*. Most notably, there is little support for the *'evolving design model'* and hence *'design for re-use'*.

Supporting *'design for re-use'* requires that we identify and abstract possible reusable knowledge elements during design itself. *'Design for re-use'* promotes the design of artefacts which meet the current design requirements but are specifically developed with integral re-use capabilities. A review of computational support for design re-use (see Fig. 9) [5] also highlighted a lack of provision for the process of *'design for re-use'* confirming that the structuring field is not alone.

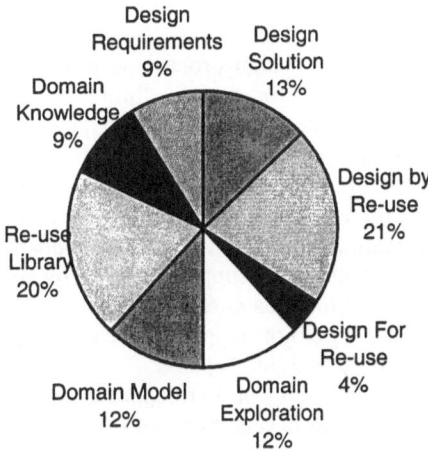

Fig. 9. Overall Computational Support

Figure 9 shows the overall support for an overall re-use approach afforded by current computational based systems. Here, 'design for re-use' was also shown to receive the least attention from supporting systems whereas 'design by re-use and the re-use library receive most effort. The study showed that 'despite the considerable number of re-use systems, there was a lack of overall process and theoretical foundation'[5] which again is similar to the findings of the product structuring mapping process discussed.

The difficulties, for product structuring practitioners, arise from the requirements of supporting continually evolving knowledge with the specific intent of creating re-usable knowledge sources. As detailed earlier (section 2.2) the motivation for recording early abstract design knowledge is confined to individual designers. As illustrated in Fig. 3 'design re-use' is more effectively facilitated when knowledge is constantly abstracted and structured for re-use. Yu [25] supports this, stating that the difficulties in creating product structures are due to a lack of product information and designing knowledge at an abstract level. Hence, to support *'design for re-use'* product structuring principles must identify and understand how the elements of a design combine and relate through all phases of the design process to result in the final solution. Here, Hansen [26] argues that although the elements of a machine system are fairly well known, the relation and their characteristics are only partly known and vaguely identified. Thus, the difficulties of structuring *'designs for re-use'* are two fold in that a) we must have initial access to all the knowledge, including knowledge that is more abstract and high level, from an evolving process and b) we must structure this knowledge effectively to support and enhance it's ability to be re-used.

The following section discusses the potential for product structuring principles to overcome these difficulties. Possible applications of enhanced structuring theories are discussed with respect to their ability to support an evolving design with the specific intent of utilising the design knowledge for re-use.

4 Structuring Potential in Design for Re-use

'*Design for re-use*', which is carried out during the design process, is primarily concerned with knowledge extraction and capture for use in future design scenarios. Generally '*design for re-use*' relates to the definition, application and management of structures in which complex design related knowledge, from an evolving design and design process, can be captured for re-use. The process draws from this '*evolving design model*' to support the construction and maintenance of a re-use library and, through the process '*domain exploration*', a '*domain model*'. The two knowledge components specifically required to support '*design by re-use*'. Inherent support for '*design for re-use*' is limited as currently 'most design environments are not evolutionary, in that they do not support the capture of newly generated knowledge and its re-use in future designs' [4]

'Brainstorming with industry highlighted the need for two approaches for storing designs for re-use (a) storing already existing past designs into the design re-use system and (b) capturing design information into the design re-use system whilst a design is being carried out' [27]. As detailed in section 3.2 the area of product structuring carries out the initial approach (a. above), with a convincing degree of success and satisfies elements of the processes of '*design by re-use*' and '*domain exploration*'. We see that, based on past designs, structuring principles can competently decompose, rationalise and configure. To achieve the latter, approach (b. above) knowledge cannot just be stored as an unstructured repository as this reduces it's effectiveness and potential application to a new design. Structuring of these repositories of knowledge is required for efficient storage, retrieval for re-use and flexibility in re-use applications. Therefore, '*design for re-use*' requires that structuring principles are made applicable to the evolution of a design.

Andreasen [16] extends the principles of the domain theory to include what is defined as functionally orientated and product life phase structuring. 'Functionally orientated structuring' relates to products' primary purpose and use while 'product life phase orientated structuring' relates to the needs of stakeholders through life phase viewpoints i.e. ease of manufacture, assembly, packaging, disposal, etc [16].

Structuring for product life phases involves 'the optimisation of properties of a design' [21] with reference to a specified viewpoint i.e. defines Design For X (where X is the viewpoint). The 'X' viewpoint is superimposed onto the product structure, the structuring orientation of 'X', may result in changes in some domains but not necessarily in others [16]. Several different structuring principles can be superimposed on a design's structure [12] as shown schematically in Fig.10.

The authors propose that the application of this 'design for x' concept be extended to include the viewpoint of 're-use'. Hence, a structure supporting re-use is superimposed onto the initial product structure to make it inherently re-usable. Thus, supporting knowledge evolution while ensuring the re-usability of design knowledge in future design scenario's. Structuring for re-use optimises utilisation of all resources obtained through and from the design process. This requires that we understand and can explicitly define elements which contribute to the re-usability of knowledge.

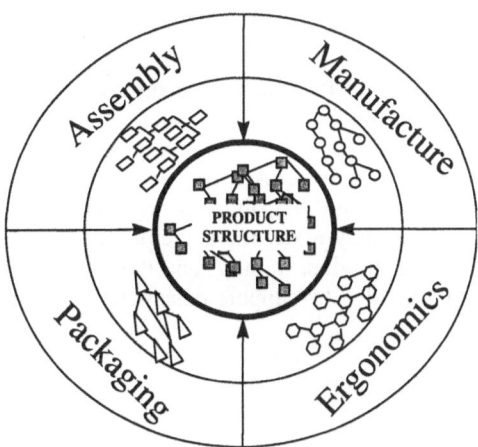

Fig. 10. Multiple Viewpoint Product Structure

For example, modularisation is one method through which *'design for re-use'* can be facilitated. Modular design involves 'the creation of a box building blocks' (including a minimum of articles) [28] which can 'fulfil various overall functions through the combination of distinct building blocks' [29]. The motivations to modularise include: the reduction of variance [16]; a trade of between flexibility and change-ability [23]; an improved proportionality of costs with respect to functionality [30]; and the provision of as an effective mechanism to upgrade and re-use existing functions, modules and assemblies [23].

Structuring principles of modularisation are embodied in work such as Erixon's Modular Function Deployment (MFD) [31], Nilsson's [32] MFD chart, Gu's Integrated Modular Design Methodology [33] and work by Elgard and Miller [34], [35].

MFD works on the principle of module drivers (the driving forces which spur the creation of modularised products) and works on the basis of identifying likely modules to support these drivers. Gu's methodology utilises clustering algorithms to identify modules based on a particular viewpoint i.e. assembly, maintenance. Elgard and Miller focus on modular engineering, a comprehensive company wide philosophy [36], in the production plant field.

The common goals for modular structuring principles are the creation of variety, reduction of complexity, and maximisation of kinship [36]. Differences lie in the modularisation focus or module driver. Some principles focus on creation of variation and reduction of complexity through the production process. While others focus variation at assembly or a reduction of complexity at a maintenance level or increased capabilities to upgrade. Defined modules can be utilised to create variants at a number of levels i.e production (same process, different size variants), assembly (same modules, different configurations). Hence, the modules are designed to support their re-use in the creation of a number of product variants. The module drivers or the focus for modularisation are the main differences between each principle. The modular principle currently supports '*design for re-use*' by identifying processes, modules and functions that can be re-used across a distinct product family.

Blackenfelt [29] states that a module 'should be based on a high degree of commonality, both momentarily and over time, to satisfy stakeholder demands during the product life phases. If this theory was extended to encompass 're-use', or the criteria for module drivers were orientated for re-use, modularisation could be seen to facilitate extended '*design for re-use*' capabilities. Extending the principles of modularity to encompass re-use may allow the principle to support not only the creation of variants across a product family but the continued utilisation and enhancement of defined modules to support design re-use over generations of product families.

5 Conclusions

The principles of 'design re-use' are cited as a worthwhile solution to meeting the requirements of a design process constrained by increasing demands in terms of quality, time, predictability and costs. The need to support knowledge through the design process is addressed and product structuring is identified as a means to achieve this. Structuring principles are mapped onto the 'design re-use process model' which is the basis for further re-use research at the CAD centre, University of Strathclyde. Discrepancies in terms of the structuring requirements of a re-use approach and current support for this approach are highlighted. Current structuring principles are shown to provide inadequate support for the process of '*design for re-use*'. The problems in supporting '*design for re-use*' are not limited to the field of product structuring and studies have shown that overall this is the least supported process of re-use.

The difficulty in supporting '*design for re-use*' is that the process requires to be carried out during design itself when knowledge, often of an abstract form, is in a continual state of evolution. It is shown that effective utilisation of this knowledge for re-use requires that it is continually supported, extracted and enhanced through structures which can support it's prolonged re-use. Thus, product structuring is highlighted as an essential element to satisfy the requirements of '*design for re-use*'.

The potential for extending the 'design for X' structuring principle to encompass the phase of 're-use' is explored. Modularisation principles, which already possess limited potential to support '*design for re-use*', are provided as an example.

The proposal raises a number of issues, including:
- How best can we extend or modify current structuring capabilities in support of '*design for re-use*'?
- How can knowledge in a continual state of flux and evolution be supported and structured?
- What structures best support knowledge utilisation to meet design requirements which are currently unrealised?
- How we can identify criteria to allow us to determine the inherent re-usability of knowledge?

6   References

[1]   Duffy, S.M., A.H.B. Duffy, and K.J. MacCallum. *A Design Reuse Model.* in *International Conference On Engineering Design.* 1995. Praha.

[2]   Sivaloganathan, S. *Foreword.* in *Design Reuse – Engineering Design Conference.* 1998. Brunel University, UK.

[3]   Culley, S.J. *Design Re-use of Standard Parts – Keynote Paper.* in *Design Reuse – Engineering Design Conference.* 1998. Brunel University, UK.

[4]   Finger, S. *Design Reuse and Design Research – Keynote Paper.* in *Design Reuse – Engineering Design Conference.* 1998. Brunel University, UK.

[5]   Duffy, A.H.B., J.S. Smith, and S.M. Duffy. *Design Reuse Research: a computational perspective – Keynote Paper.* in *Design Reuse – Engineering Design Conference.* 1998. Brunel University, UK.

[6]   Gao, Y., I. Zeid, and T. Bardasz, *Characteristics of an effective design plan system to support reuse in case-based mechanical design.* Knowledge Based Systems, 1998. 10: p. 337-350.

[7]   Khadilkar, D.V. and L.A. Stauffer, *An experimental evaluation of design information reuse during conceptual design.* Journal of Engineering Design, 1996. 7(4): p. 331-339.

[8]   Jones, M.E., *Reusing integrated circuit designs.* Computer Design, 1995. July.

[9]   Sreeram, R.T. and P. Chawdry. *A design reuse model for collaborative product development process.* in *Design Reuse – Engineering Design Conference.* 1998. Brunel University, UK.

[10] Duffy, A.H.B. and A.F. Ferns. *An Analysis of Design Reuse Benefits.* in *International Conference on Engineering Design.* 1999. Munich.

[11] Cross, N., *Engineering Design Methods: Strategies for Product Development.* 2nd ed. 1998: Wiley & Sons.

[12] Andreasen, M.M., A. Duffy, and N.H. Mortensen. *Relation Types in Machine Systems.* in *WDK Workshop on Product Structuring.* 1995. Delft University of Technology, Delft, The Netherlands.

[13] WDK. *Final Day Discussion.* in *WDK Workshop on Product Structuring.* 1995. Delft University of Technology, Delft, The Netherlands.

[14] Duffy, A.H.B. and S. Legler. *A methodology for structuring past designs.* in *4th WDK Workshop on Product Structuring.* 1998. Delft University of Technology, Delft, The Netherlands.

[15] Duffy, A.H.B. and S.M. Duffy, *Learning for Design Reuse.* Artificial Intelligence in Engineering Design, Analysis and Manufacturing, 1996. 10: p. 139-142.

[16] Andreasen, M.M., C.T. Hansen, and N.H. Mortensen. *The Structuring of Products and Product Programmes.* in *2nd WDK Worshop on Product Structuring.* 1996. Delft University of Technology, Delft, The Netherlands.

[17] Erens, F. and K. Vershulst. *Managing System Design.* in *WDK Workshop on Product Structuring.* 1995. Delft University of Technology, Delft, The Netherlands.

[18] Hansen, C.T. *An Approach to Configuring Mechanical Systems.* in *WDK Workshop on Product Structuring.* 1995. Delft University of Technology, Delft, The Netherlands.

[19] Liedholm, U. *Conceptual Design of Products and Product Families.* in *4th WDK Workshop on Product Structuring.* 1998. Delft University of Technology, Delft, The Netherlands.

[20] Yu, B. and K.J. MacCallum. *A Product Structure Methodology to Support Configuration Design.* in *WDK Workshop on Product Structuring.* 1995. Delft University of Technology, Delft, The Netherlands.

[21] Tichem and T. Storm. *Issues in Product Structuring.* in *2nd WDK Worshop on Product Structuring.* 1996. Delft University of Technology, Delft, The Netherlands.

[22] Herbertsson, J. *Product Structuring in Design for Manufacture.* in *WDK Workshop on Product Structuring.* 1995. Delft University of Technology, Delft, The Netherlands.

[23] Erens, F. and K. Vershulst. *Architectures for Product Families.* in *2nd WDK Worshop on Product Structuring.* 1996. Delft University of Technology, Delft, The Netherlands.

[24] Elgard, P. *Industrial Practices with product Platforms in USA.* in *4th WDK Workshop on Product Structuring.* 1998. Delft University of Technology, Delft, The Netherlands.

[25] Yu, B. and K.J. MacCallum. *Product Structuring in Reality.* in *WDK2 Workshop on Product Structuring.* 1996. Delft University of Technology, Delft, The Netherlands.

[26] Hansen, C.T. *Towards a Tool for Computer Supported Configuring of Machine Systems.* in *2nd WDK Worshop on Product Structuring.* 1996. Delft University of Technology, Delft, The Netherlands.

[27] Shanin, T.M.M., P.T.J. Andrews, and S. Sivaloganathan. *A design reuse system.* in *Design Reuse – Engineering Design Conference.* 1998. Brunel University, UK.

[28] Janson, L. *Business Orientated Product Structures.* in *WDK Workshop on Product Structuring.* 1995. Delft University of Technology, Delft, The Netherlands.

[29] Blackenfelt, M. and S. Andersson. *Modularity by Distribution – the development of a mobile robot prototype platform.* in *4th WDK Workshop on Product Structuring.* 1998. Delft University of Technology, Delft, The Netherlands.

[30] Pels, H.J. *Modularity in Product Design.* in *3rd WDK Worshop on Product Structuring.* 1997. Delft University of Technology, Delft, The Netherlands.

[31] Erixon, G. *Modular Function Deployment.* in *2nd WDK Worshop on Product Structuring.* 1996. Delft University of Technology, Delft, The Netherlands.

[32] Nilsson, P. and G. Erixon. *The Chart of Modular Function Deployment.* in *4th WDK Workshop on Product Structuring.* 1998. Delft University of Technology, Delft, The Netherlands.

[33] Gu, P., M. Hashemian, and S. Sosale, *An Integrated Modular Design Methodology for Life-Cycle Engineering.* CIRP Annals, 1997. 46 (1): p. 71-74.

[34] Elgard, P. and T.D. Miller. *Designing Product Families.* in *Design for Integration in Manufacturing. Proceedings of the 13th IPS Research Seminar.* 1998. Aalborg University, Fugsloe.

[35] Miller, T. and P. Elgard. *Defining Modules, Modularity and Modularisation – evolution of the concept in a historical perspective.* in *Design for Integration in Manufacturing. Proceedings of the 13th IPS Research Seminar.* 1998. Aalborg University, Fugsloe.

[36] Riitahuhta, A. and M.M. Andreasen. *Metrics for supporting the use of Modularisation in IPD.* in *4th WDK Workshop on Product Structuring.* 1998. Delft University of Technology, Delft, The Netherlands.

Capturing Quality Perceptions in the Design Rational of a Modular Product Concept

Mark W. Lange, Johan Åslund

Royal Institute of Technology
Department of Computer Systems for Design and Manufacturing
Balmelvägen 68
SE-100 44, Stockholm
e-mails: mark.lange@reachin.se, johan.aslund@modular.management.se

Abstract. The design of a product is a process of negotiating product knowledge represented in two general forms, 1) verbal and textual statements in a "design rational" about a product and 2) models of a product using a graphical modeling languages. The objective of this process is an artifact with which a user will interact, an interaction based on the use of our senses. The problem here is that, when considering the use of computer systems to support communication during integrated product development, the rational behind a design, representing what is understood about a product lacks a quantifiable relationship to the models of a product. An approach to understanding this problem can be found in the domain of semiotic sign theory, where product synthesis in design is treated as an act of semiosis. This approach is discussed in this paper in conjunction with product structuring supported by the method, Modular Function Deployment, the result of which is presented as a portion of a general theory of synthesis called Design Semiosis.

1 Introduction

Design has been described as a process of transforming of a set of stated needs into an artifact [12]. This process of transformation contains a number of general activities, described by Andreasen in [1] as,

- *problem solving*, i.e. the activities associated with creating solutions with specific aims in mind;
- *product synthesis*, i.e. the activity of creating a specific product from the formulation of a task;
- *product development*, i.e. the activity of creating a product-based commercial activity (including production and marketing) on the basis of a recognized need or a contract from a customer;
- *product planning*, i.e. the activities of implementing and coordinating strategies for the product, market and technology, within a range of products which cover the needs of the market and yield the necessary profits.

These different activities have received a great deal of attention in design research, with the possible exception of the activity of *product synthesis*. Approaches to synthesis have been through "design by analysis" [15]. In this sense, designers decompose one level of product knowledge units into a lower level of product knowledge units, following stringent rules, such as the "independence" and "information" axioms found in axiomatic design [14]. As such, synthesis in design research is approached from a viewpoint of an act of logic.

Fig. 1. A variety of solutions designed from a common need

Design, particularly the synthesis of a product, is an activity unique to humans, where no two designers produce the same result, not even when driven logically by a common need. This is exemplified by the proliferation of different mobile telephone designs on the market.

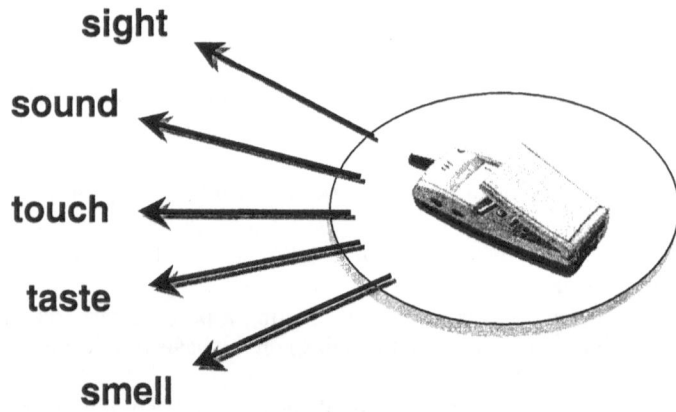

Fig. 2. All the senses are used to experience the intended "effect" of a designed product.

A product is experienced through an interaction with the five senses; sight, sound, touch, taste and smell. It is through these senses and the effects of understanding they engender that a designer must consider when designing a product. Yet, products are designed nearly exclusively through the sense of sight using predominantly graphical representations.

An alternative approach to the understanding of synthesis can be found by treating design as a semiotic act using signs. This enables the application of semiotics, the science of signs and sign theory [8], to the problem of understanding synthesis. With a semiotic viewpoint, the common denominator is neither the product itself nor the stakeholder in the product, but in a relationship containing both *and* the meaning that is made with the relationship of these two aspects. This is captured in the concept of the semiotic element, a 'sign', which, through the use of our senses, we read in the products that we design.

The focus of the research that is reported in this article is the classification of the indication of quality perceptions in the structuring of a modular product concept. The article will continue with a presentation of a foundation for a set of theories, which describe a research approach to this problem area called *design semiosis* [6]. These theories will then be set in a case study clarifying development of a design rational using Modular Function Deployment [2], a method used for the development of modular product concepts. Finally, a discussion will be presented on future issues of these results, both industrially and theoretically.

Logic based approaches to synthesis in design only treat the decisions made during design. But this method for design does not clarify how products are designed to please the stakeholder or why there are different solutions to products generated by designers, when the same or similar sets of requirements are used in the design effort. From a research point of view, this begs the question, 'is there a basis for design that emphasizes the means through which products are appreciated?'

2 Foundations

2.1
Design Semiosis and Its Sign Theory

In semiotics, a sign is a semiotic entity produced through the act of giving meaning to something by someone [11]. In this sense a sign is not a physical signal, a sign element or a sign vehicle; these are all sources of stimuli, signals that strike our senses. It is to these stimuli that we give meanings; these stimuli do not have any meaning outside of their relation to us.

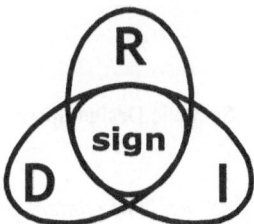

Fig. 3. Design Semiosis 'Sign' Symbol [6]

This symbol represents a semiotic sign that occurs when there is an established relationship between a desired effect in an Interpretant, a label or identity as a

Designant and a vehicle for the sign as a Representamen. Without all of these constituent aspects in place there is no sign that has meaning [6].

It is through our senses that we interact with our environment. We hear, see, touch, taste and smell all the artifacts that we use in our daily routines. Through our understanding of the stimuli on our senses, or 'signs in the mind' (an interpretant), we give 'sign meanings' (a desginant) to these 'sign vehicles' (the representamen) [6]. Together these three qualitative aspects form what is called, in semiotics, the sign triad.

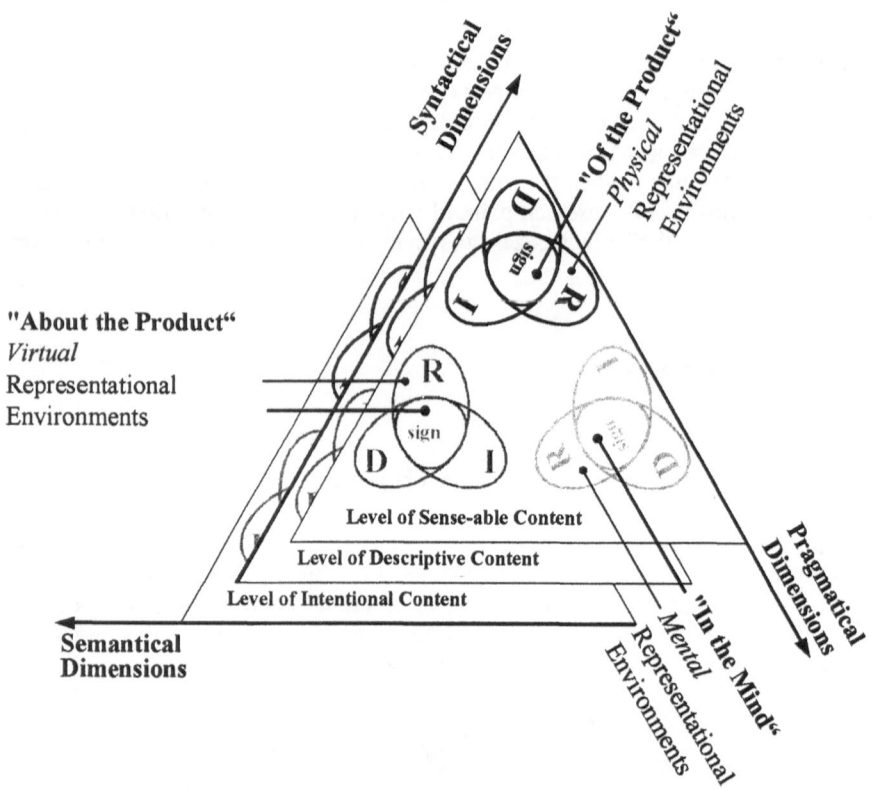

Fig. 4. The Levels of Content in a Semiotic Design Space

Three qualitative dimensions and levels of content prescribe the semiotic space of product design. In turn, each qualitative dimension has a quantitative aspect that corresponds to the levels of content. Using this model, the information content of design synthesis can be mapped through a variety of concatenated and transformed semiotic models [6].

This sign, though, is not a static relationship; it is a dynamic entity and as such also has quantitative levels of content. These levels of content, in turn reflect the qualitative aspects of the sign triad; sense-able content corresponds to the representanmen, descriptive content corresponds to the designant and the intentional content corresponds to the interpretant. In the design of a product, representations "of" the product, as product models, are found in the graphical languages; Sketches (2D), diagrams (2½D), feature-based solid models (3D) from the basis of this way of speaking "of" the product. These ways of speaking "of" a product follow a range of semiotic content levels of icons, indexes and symbols. Representations "about" the product are found in the design rational using natural languages. These ways of speaking "about" a product follow a range of semiotic content levels of quality, designant and intrinsic properties. The following sections contain definitions of these properties as they function in design semiosis.

2.2
The Content of a Design Sign

The first part of a sign triad addresses the syntactical aspect of the sign and is called the *representamen* and is the foundation of the sign triad. This is the form that the sign takes and serves as its "vehicle". Representamen's, as sign vehicles, are used during design to represent a meaning regarding some attribute of a solution being generated and have three general forms described as follows,

• **Icons** – Literally resemble the attribute, or attributes, of an object that they portray. An example is a digital feature-based solid model projected in a virtual reality display system or the form and geometric features used to construct the feature-based solid model, such as drafted surfaces, radii or chamfers.

• **Indexes** – Initiate a resemblance to the attributes of an object through causation. Examples of indexes are the "click" of a micro-switch, the "hum" of a vacuum cleaner. In graphical modeling the use of view projection is a case of an index.

• **Symbols** – Relate to the attributes of an object through convention. The letters of the alphabet, mathematical operators and function sketches are examples of this type of designant.

Representations of a product during its design move through transformation from an abstract state of representation, using symbols, to a concrete state of representation, using icons.

The semantical aspect of the sign, referred to as the *designant*, addresses the second part of the sign triad. This is the identity, or label, engendered by the representation, of the sign. In other words, this is what it "stands for". Aspects of this portion of the sign triad also have three levels of engendering,

• **Quality Properties** – These vehicles are directly sense-able and represent such designed properties such as colours, form, textures, sounds, etc.

• **Designate Properties** – These are properties of a product that have a dynamic nature. They are not directly sense-able, yet occur as a part of our collective reality, such as steel, wood, plastic, sound density, velocities, accelerations, functions, actions, etc.

- **Intrinsic Properties** – Properties that are driven by the physical laws of nature. Examples of this are Coulomb's Law, Newton First Law and the coefficient of friction.

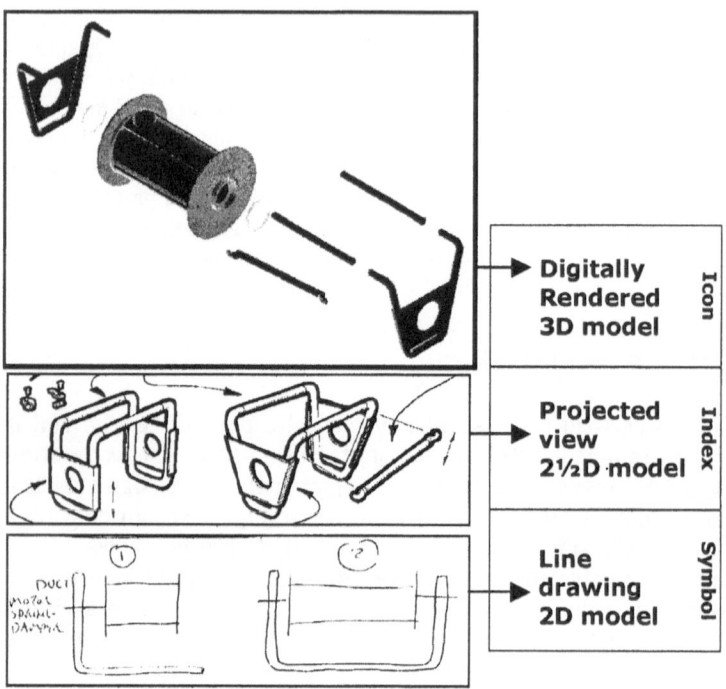

Fig. 5. Progression of the Quantitative Forms of the Representamen (The Sign Vehicle) [5]

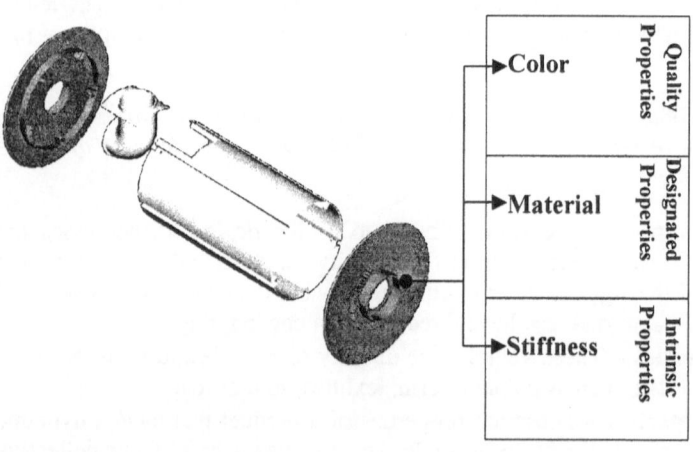

Fig.6. Quantitative Forms of the Designant : The Sign Vehicle [5]

The meanings given to sign vehicles generally fall within three levels of content, the Quality, Designate and Intrinsic Properties [5].

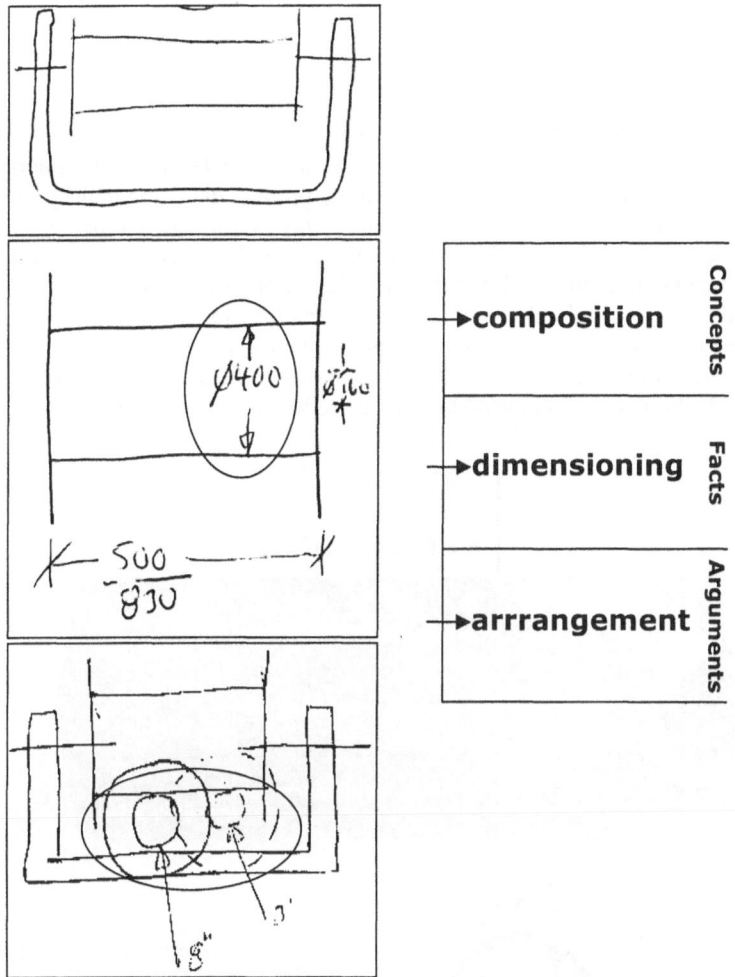

Fig. 7. Quantitative States of the Interpretant : The Sign of the Mind [5]

The effect engendered by a representation follow three general states of understanding, the state of a concept, fact or argument [5].

The triad is completed with the pragmatic aspect, or the *interpretant*. This is the sense, or understanding, made of the sign. This sense itself takes the form of a "sign of the mind". Note here that the interpretant is not the "interpreter" of a sign, the interpretant is the "effect of understanding" produced by the sign. This "effect of understanding" is also the basis of the creation of additional signs.

- **Concepts** – These are possibilities, ideas, and effects of understanding a solution.
- **Facts** – These are descriptive annotations reflecting a distinct unit of knowledge.
- **Arguments** – These are reasons or propositions that require an understanding.

2.3
Products as Sign Carriers

From the domain of industrial design, Monö states that products are the carriers of, *"...messages in a 'language' that we see, hear or feel...[7]."* Designers communicate through the signs that they place in the product using the concept of a 'current product sign'; *"The market's conception of the way in which a product's principal function is traditionally presented in its gestalt [7]."* The gestalt of a product is the means through which we perceive the quality of a product through its gestalt. Take for example the proliferation of mobile telephones currently available on the market; each one has its own gestalt and in each case this gestalt is perceived using our senses. The properties of the product that are to be perceived must be known and indicated in a design rational.

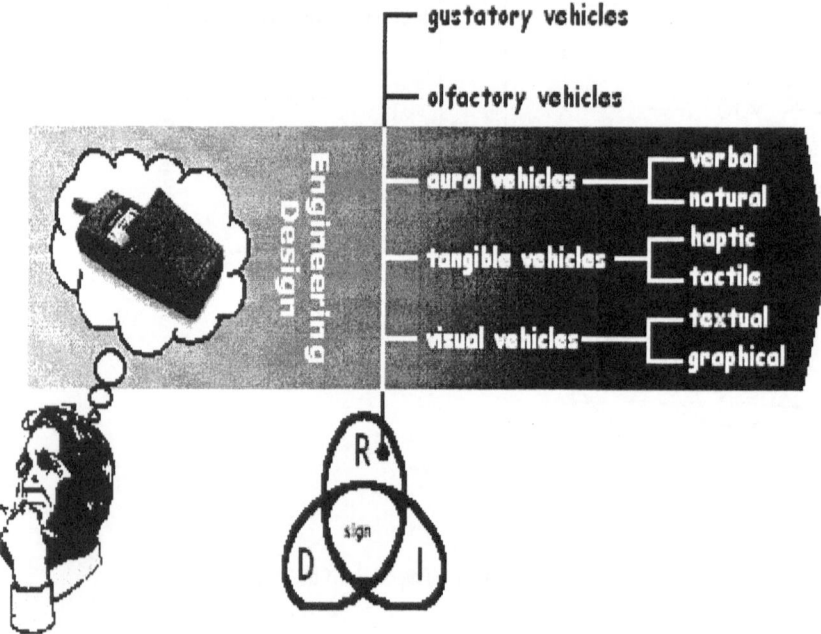

Fig.8. The Sign Vehicle Groups

There are five main sign vehicle groups that can be utilized in the representation of information in a design solution for a product; visual, tangible, auditory, olfactory and gustatory. Each group of vehicles corresponds to a different sense through which the information about the product shall be understood.

This viewpoint on product design is important because it is the basis of perception; we use our senses to interact with our environment. We hear, see, touch, taste and smell artifacts to which we give meanings, as we understand them. These interactions occur at every possible phase of a product's life, beginning with an idea in the mind of an entrepreneur and ending with the product in the hands of a recycler. It is through our senses that we establish our perception of the quality of a product.

2.4
Quality through Perception

Mørup states that,
   "...quality is experienced when the customer interacts with the product..."
   [9, p. 90]

Interaction with a product requires our senses. It is through our senses that we perceive the quality of a product. As for customers, there are different types and each has a different view on the quality in a product. Mørup makes a distinction between two different types of stakeholders when regarding the definition of quality, *external* (e.g. customers, end users, approving authorities and external sales and service people) and *internal* (e.g. all internal functional areas and employees in contact with the product such as designers, production, quality control, etc.).

From this division of stakeholders Mørup defines two concepts for quality; *Q-quality*: Q is the customer's qualitative perception of the product and *q-quality*: q is the internal stakeholder's qualitative perception of the product in relation to their product-related tasks.

Following these definitions, everyone having contact with a product has a perception of its quality through its quality properties. For the internal stakeholders perceptions follow the life cycle of the product. Beginning with and idea in the mind of an entrepreneur, ending in the hands of a material recycler, and every person in between, each has a perception of the product properties from the perspective of how they interact with it.

2.5
Product structuring

Modularization of products has been shown to be a successful product development, particularly in Swedish industry [3, 4]. Based on empirical research in Swedish industry, the reasons for modules in a product can be described according to a set of Module Drivers [16]: A method has been developed to support modular product development called Modular Function Deployment (MFD) [2]. Considering this success of the method, it has been accepted that these results are significantly viable that further research into the domain of modular product design and development can be based on Modular Function Deployment [13].

The focus, for this article, is Step 3 and the use of the Module Indication Matrix (MIM) to capture and represent a set of information and data as the design rational, which is a graphical statement of the structure of a modular product. The objective of the MIM is to determine the suitability of a technical solution (TS) to serve as a module foundation, according to a set of company specific module drivers (MDs). This is performed by weighting the relevance of each MD for a given TS.

3 A Study

The synthesis of a solution must begin from a known state of understanding, in this case a specification of a *product structure*. This product structure is presented in the forma of an MFD Chart [10]. This state of understanding is in turn semiotically operated upon, based on a range of other states of understanding. One of the operations performed is the categorisation of the product's properties in terms of its Quality, Designate and Intrinsic Properties. An understanding of corporate strategies, design experience, product domain and life cycle awareness controls these carburizations.

This is exemplified in the case of a mobile telephone development exercise supported by the MFD method. The resulting decisions and indications of decisions, represented as statements in the design rational, is placed in a matrix structure, which highlights the relations between three primary information categories; the Voice of the Customer as customer demands, Quality Perception as

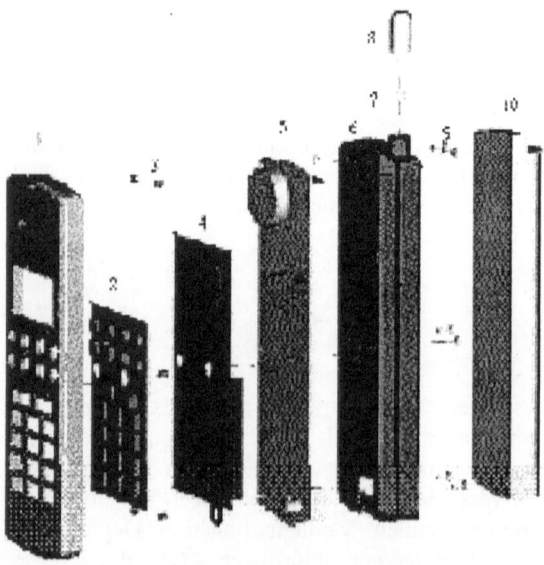

Fig. 9. Modular Telephone Model

product properties and the Means of Realization as technical solutions. The understandings of these statements support the evaluation of the technical solutions by the module drivers, which, according to the module driver profile, is used to derive a product structure.

This is an image of the digital product model for a telephone produced following the understanding of the design rational contained in the MFD Chart. Note, this image is an ICON following the theories of Design Semiosis.

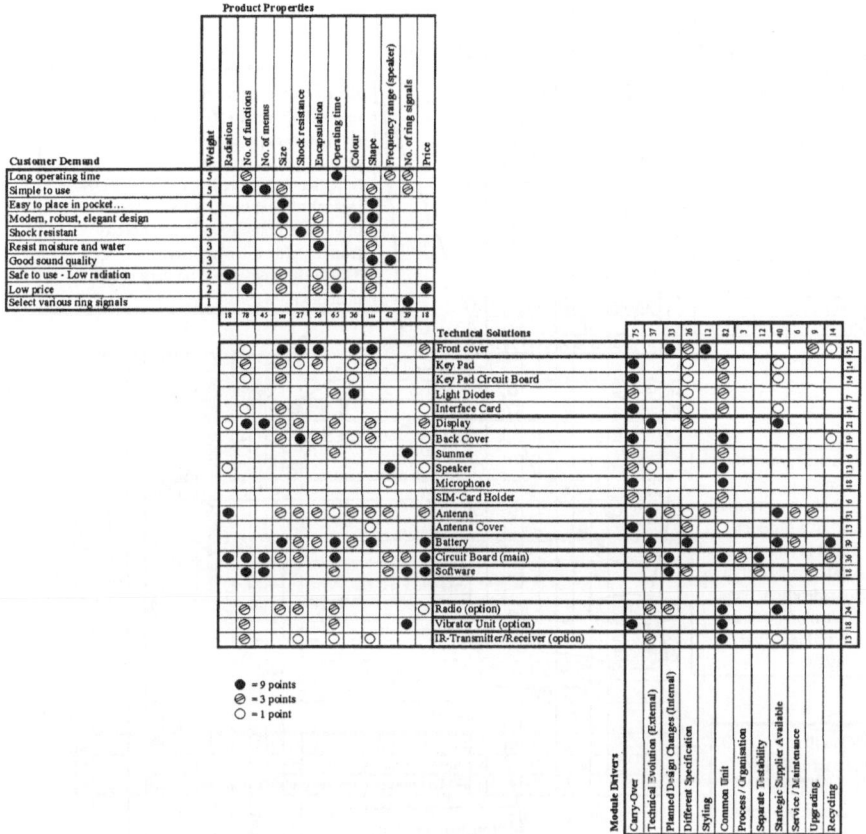

Fig.10. The Mobile Telephone Design Rational (before semiosis)

From this original "specification", the product properties have been further sorted according to the semiotic levels of content described in the theories of Design Semiosis; Quality, Designated and Intrinsic Properties: From this categorisation an analysis can be made through the connection matrix that lies between the QFD and MIM sections.

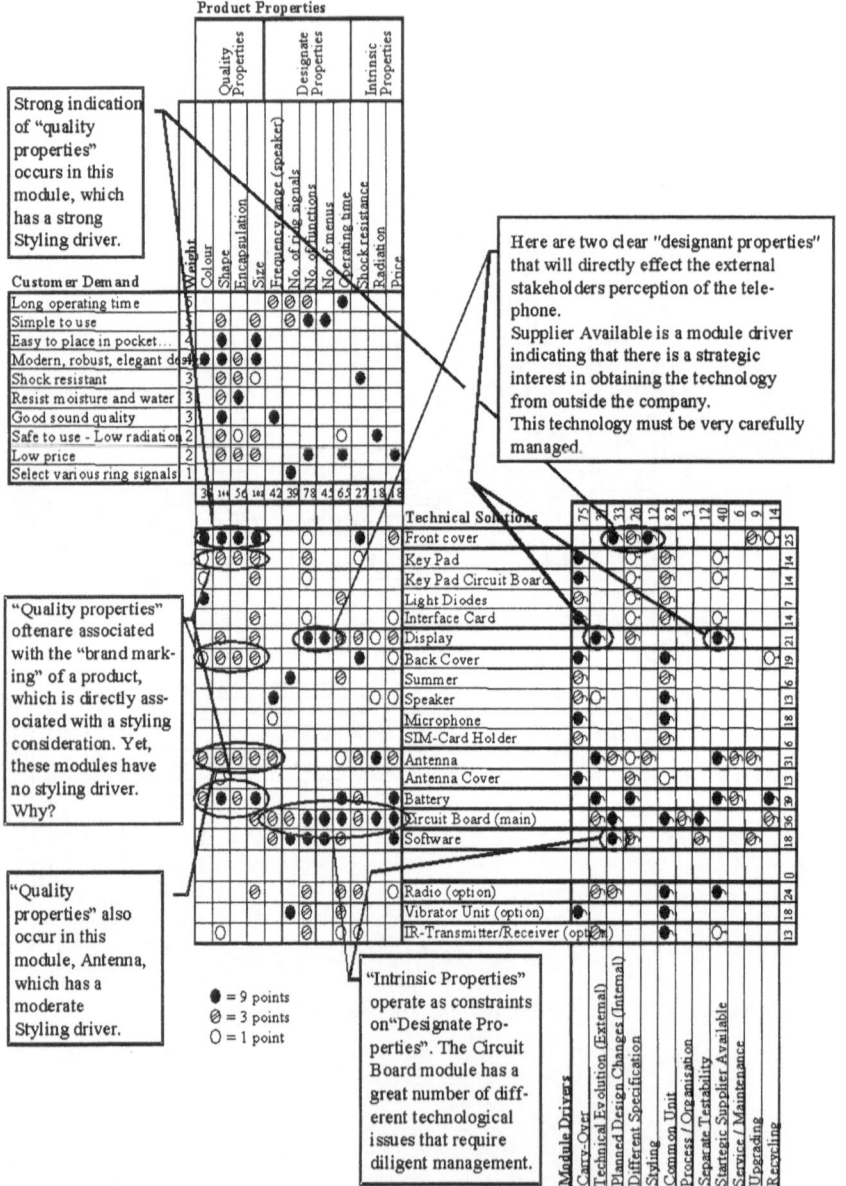

Fig. 11. The Mobile Telephone Design Rational (after semiosis)

The analysis of this training example is made from the design viewpoint of where the *brand marking* will be designed in the product. These understandings follow the semiotic levels of content in the product properties, which should be reflected in the indications of the module drivers. A review of these understandings is presented in Fig 10.

This is a summary of the design rational, presented in an MFD Chart [Nilsson & 21], textual statements represent the understandings of what the product is to contain in terms of its fulfilling "customer demands" that are enabled as "product properties", which in turn are embodied in "technical solutions" structured according to "module drivers". All this information is used to guide the synthesis of a product.

In this updated summary of the design rational the semiotic levels of the product properties is more clearly reflected in the proposed product structure. This includes some clear indications that there are quality properties that are not indicated in the module driver profile (key pad, back cover, antenna and battery modules). In addition, there are some modules, such as the circuit board, that have a great number of "Designate and Intrinsic Properties" that support the indication of a Planned Design Change driver. Note that all the modules considered in this analysis, with the exception for the Circuit Board, are not indicated as "Common Units".

4 Discussion

This paper has briefly introduced the application of a very small portion of a semiotic sign theory, the basis of understanding design semiosis. This article has presented a view that the design rational placed in a structured format and represented as a graphical statement, can be categorized according to quantitative levels of semiotic meaning. This based on the design of products that are the carriers of signs that engender meaning through our senses. In a broader sense, this semiotic viewpoint has many implications on the domain of engineering design and integrated product development. These implications are namely,

4.1
Design and Engineering Education

Design Semiosis requires a variety of different inputs, controls and supports in order that a solution can be synthesized.

An understanding of sign semiosis can serve as a "vehicle" for teaching a "quality mind set" in engineers and designers. Semiosis concerns not just the gestalt of a product, having a representational form, but also concerns what different representational forms (of the product) have for meaning (about the product) to us as designers and engineers.

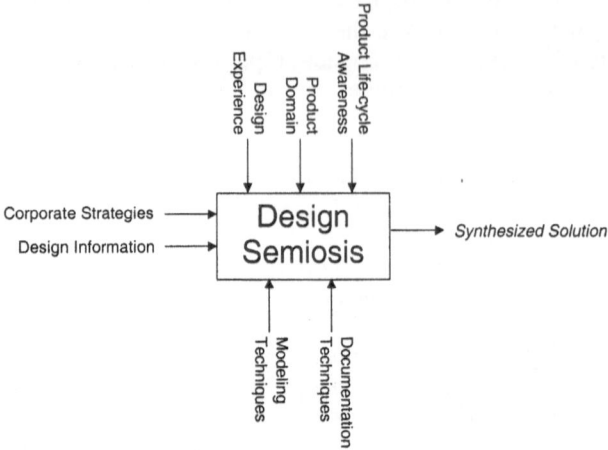

Fig.12. The Activity of Synthesizing a Solution

4.2
Brand Marking and Product Development Strategies

In multi-national companies, the challenge is to keep as many of the technical systems common between the different products in the platform, without loosing the brand identity in each product. For example, an Audi must be an Audi; despite the fact that majority of its components will be shared with other VAG cars in the future. The need to understand what in a car is brand unique and what can be shared with other models rises each time a cross brand car platform is developed.

4.3
Virtual Reality

The application of technologies related to virtual reality enables the presentation of a product that affects more than just the sense of sight. Using computer driven graphical interfaces integrated with haptic, tactile, and aural sense-ability, designers and customers can interact with product designs enabling an earlier understanding of the product "quality properties".

5 Conclusion

In order to use a product we must interact with it. This interaction with the product is achieved through our senses of touch, taste, sight, sound and smell and enables us to perceive quality. Yet, the design of a product is not driven by the understanding of where quality is perceived, but from the need to manage the development of the technological content. During the early the phases of the product life cycle it is most critical that the perception of the product quality is represented in the design rational, considering that decisions made early in the product life have a significant impact later in the product life. This paper has described a small portion of a broader research focus on synthesis in design called Design Semiosis, with a clarification of how quantitative levels of sign semiosis, in the form of product properties, are identified in the structure of the design rational for a modular product concept.

6 Acknowledgements

This research has been financially supported in part by ReachIn Technologies AB, www.reachin.se and in part by the Swedish Foundation for Strategic Research, www.stratresearch.se/welc-e.htm through the Swedish Engineering Design Research and Education Agenda, ENDREA, www.endrea.sunet.se. The authors would also like to acknowledge Modular Management AB for the kind use of the MFD training material used as an exemplification in this article.

7 References

[1]   M.M. Andreasen, "Design Methodology", Journal of Engineering Design, Oxford, Vol. 2, No. 4, 1991.

[2]   G. Erixon, Modular Function Deployment – A Method for Product Modularization, Doctoral Thesis, The Royal Institute Of Technology, Dept. of Manufacturing Systems, Assembly Systems Division, Stockholm, Sweden, 1998.

[3]   G. Erixon, J. Fredriksson, L. Romson, and A. von Yxkull, Modulindelning I Praktiken, Sveriges Verkstadsindustrier (in Swedish only), ISBN 91-7548-460-9, 1996.

[4]   G. Erixon, A. Erlandson, A. von Yxkull and B.M. Östgren, Modulindela produkten - halvera ledtider och offensiv marknadsorientering. Industrilitterature (in Swedish only), ISBN 91-7548-321-1, 1994.

[5]   M.W. Lange, "CONCAD Bridging; Presenting a Theory for Studying Design as Sign Semiosis". Submission to the American Society of Mechanical Engineers, 12th Conference on Design Theory and Methodology. 2000.

[6]   M.W. Lange, CONCAD Bridging; Supporting communication between Concept and Detail Design Activities in Product Development and Design, Licentiate Thesis (under preparation), Div. Computer Systems for Design and Manufacturing, Dept. Production Systems, Royal Institute of Technology, Stockholm, Sweden, 2000.

[7]   R. Monö, Design for Product Understanding: The Aesthetics of Design from a Semiotic Approach, Stockholm, Sweden: Liber AB, ISBN 91-47-01105-X, 1997.

[8]   C.W. Morris, Signification and Significance: A Study of the Relations of Signs and Values, Cambrige, MA: The MIT Press, ISBN 02-62630-14-1, 1964.

[9]   M. Mørup, Design for Quality, Doctoral Thesis, Institute for Engineering Design, Technical University of Denmark, Lyngby, 1993.

[10] Nilsson, P. and Erixon, G.; The Chart of Modular Function Deployment; 4th WDK Workshop on Product Structuring, October 22-23, Delft University of Technology, The Netherlands; 1998.

[11] C.S. Peirce, The Collected Papers of Charles Sanders Peirce, C. Hartshome, P. Weiss and A.W. Burks (eds.), Cambridge, MA; Harvard University Press, 1931-58.

[12] P. Sargent, S. Konda, I Monarch and E. Subrahmanian, "Shared Memory in Design: A Unifying Theme for Research and Practice", Research in Engineering Design, Vol. 4, No. 1, pp. 23-42, 1991.

[13] Stake, R.; A hierarchical classification of the reasons for dividing products into modules; A theoretical analysis of module drivers; Licentiate Thesis, Div. Of Assembly Systems, Dept. of Manufacturing Systems, Royal Institute of Technology, Stockholm, Sweden; 1999.

[14] N.P. Suh , *Principles of Design*, Oxford University Press, 1990.

[15] M. Yoshioka and T. Tamiyama, "Toward a Reasoning Framework of Design as Synthesis", Proceedings of the ASME 1999 Design Engineering Technical Conferences, Sept. 12-15, Las Vegas, Nevada, American Society of Mechanical Engineers, NY, ISBN 0-7918-1967-1 (CDROM), 1999.

[16] B.M. Östgren, Modulindelad produkt - effekter in hela företaget (in Swedish only), Licentiate thesis, Dept. of Manufacturing Systems, Royal Institute of Technology, Stockholm, Sweden, 1994.

Relating Product and Process Structures

Alison McKay, Alan de Pennington, Steven J. Trott

University of Leeds
School of Mechanical Engineering
Leeds LS2 9JT, UK
e-mail: mckay@leva.leeds.ac.uk

Abstract. The management of design processes, and the many change processes that occur therein, requires an appreciation of the relationship between the product being designed and the processes that are creating and will use the resulting product definition. This paper reports on early experiments in which product structuring techniques are applied to product definition processes (such as design and change) with a view to creating compatible definitions of product and process. From this basis, the creation of related product and process definitions becomes a viable goal. An information framework for the unified representation of product and process is presented and its relationship to the STEP data specification architecture is outlined. The framework draws upon concepts from product data engineering technology and is demonstrated through population with product structure data and associated process definition data.

1 Introduction

The computer-based representation of product data has evolved since the early 1960's from being a 2D [CAD] representation of engineering drawings through wire frame, surface and the solid 3D representation schemes to, relatively recently, product structures that carry detailed data typically related to the properties of products. In parallel with this evolution, opinion has begun to move from a perceived need for a single representation with subsets to support different viewpoints (as in the application schemas of the ANSI/SPARC three layer model) to one that supports different product structures that carry product data specific to given tasks according to the life-cycle stage and activities that the product structure is to support. Product structuring is a means by which domain specific frameworks for people in different roles (and so with different viewpoints) can be superimposed on the detailed representations of the properties of a given product.

The need to bound product data to be captured in an engineering information system has led to the need to identify information requirements. These are typically identified through analysis of the process to be supported and are codified through the use of modelling techniques such as IDEF0 and IDEF3. The biggest manifestation of this approach is embedded within the STEP AP development methodology where AAMs (Application Activity Models) and ARMs (Application Reference Models) are developed before the exchange format, the AIM (Application Interpreted Model), is defined. This paper is based on the assertion that the techniques used yield process definitions that are analogous to the shape representations of products. Further, like product, to make sense and be usable by people who execute and manage the processes concerned, domain-specific frameworks to support different viewpoints of the same process and/or activities are needed.

A major criticism of engineering information systems is that they often do not support [are not aligned with] the engineering processes that they are supposed to facilitate/improve. Better understanding of the process and the relationship between the process and the product data that it deals with is needed. In this paper initial experiments in applying product structuring technology to the representation of processes are reported. The ultimate goal of this work is to provide sufficient understanding of relationships between product and process, be that the product definition process or the product realisation process, to allow more effective engineering information systems to be created and used. Here a case study of a process and associated product structures is presented and a way in which product structuring technology can be applied to capture the process and product structures, and relationships between them, is demonstrated.

2 Background and Related Work

Key concepts from the STEP data specification architecture [1] can be used to categorise current approaches to the representation of product, process and the relationships between them. These concepts are given down the left hand side of Table 1. Models that identify and describe without significant reference to either decomposition or relationships to other elements fall into the *Definition* category; this aligns with the STEP *product* and *product-definition* entities. More sophisticated models allow the composition and/or interfaces to other elements to expressed and fall into the *Definition and/or composition structure* and *Definition and/or connectivity structure* categories respectively; these align with different usages of the STEP *product_definition_relationship* entity. Finally, some models include details of properties and so fall into the *Property definition/representation* category; this aligns with the STEP entities that allow product properties to be defined and represented.

Andersson: 1999 [2] proposes requirement, concept and behaviour models of products being designed; each product definition relates to a given stage in the design process. Many authors propose definitions of design processes and these are beyond the scope of this paper; definitions of processes that are used to relate product details to the process are, however, within the scope of this paper. For example, Bichmaier and Grunwald [3] outline a design process definition based on four elements (analysis, synthesis, evaluation, selection) and relate product properties to processes that verify the properties. In a similar vein, Cohen, Peak, Fulton [4] discuss change processes and the causes of change across a supply chain; supply chain processes are the focus. Malmqvist and Schachinger [5] discuss product modelling in a product introduction process that is based on and corresponds to Andreasson's chromosome model [6]. Svensson, Malmstrom, Pikosz and Malmqvist [7] offer four perspectives (process, information, role, systems) and a relationship matrix to relate pairs of perspectives. Trott, Baxter, McKay, de Pennington and Henson [8] outline an approach based on making relationships between different product structures explicit and so available for queries. This paper takes this work forward in exploring relationships between the product data and the process that creates and uses it.

Other researchers focus on product and relationships to process. For example, Peng and Trappey [9] relate physical assembly structure with change processes through the notion of versions of product data. IDEF0 and IDEF3 [10, 11] are used to capture representations of processes and activities that include both composition and connectivity structures. Westfechel [13] defines a product development management process and relates versions of a product definition to it; hence this work sits in the definition and connectivity rows of Table 1.

Table 1. Definition and connectivity in relation to product and process

	Product	Process	Related product & process
Definition	Requirement, concept, behaviour [2]	Analysis, synthesis, evaluation, selection [3]	Relating product to process: design process [3], supply chain process [4], product introduction process [5]. Relationship definition format [7].
Definition and/ or composition structure	Bills of materials, Function structures [8,16]	IDEF0 [10] IDEF3 [11]	Product structure and engineering change process [9]
Definition and/ or connectivity structure	Function structures, Mating conditions [8,16]	IDEF0 (flows) [10] IDEF3 (sequence) [11]	Versions of products in a process [13]
Property definition, representation and/or presentation	Geometric models	IDEF0 & IDEF3 are candidates	STEP Industrial Data Framework [14]

The STEP Industrial Data Framework [14] defines products in the context of a product lifecycle reference model; it shows relationships between processes and product flowing and suggests that relationships between product and process are through product properties and when the product is in a given state. The idea that products are frequently related to processes through their properties aligns with the findings of the authors that are reported in [18]. The example that is used in this paper relates process structure in terms of its connectivity with product composition structures, all at the definition level of Table 1.

3 Case Study

The product selected as a case study for this paper is the nose cowl of Rolls Royce's Trent 700 aero engine nacelle. The case study considers a Product Introduction Process (PIP) across two tiers of a supply chain. The Rolls-Royce Integrated Product Development (IPD) Process [15] (see Fig. 1) was used to scope the process.

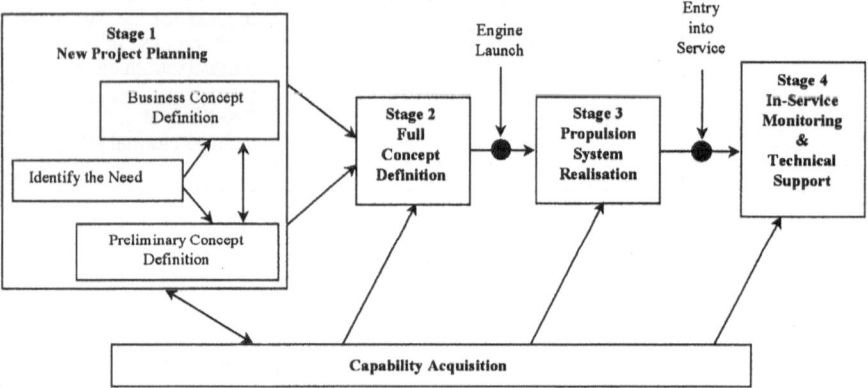

Fig. 1. Rolls-Royce Integrated Product Development (IPD) process

The Rolls Royce IPD process supports a product lifecycle from new project planning, through full concept definition and system realisation, to in-service monitoring and technical support. It is composed of the essential activities that together achieve the delivery of fully verified propulsion systems that satisfy customers' needs. This includes the delivery of the propulsion system as an engine ready for installation and a representation of the propulsion system that may be regarded both as an integrated whole and as a major functional unit of the propulsion system.

The IPD process is used as a general purpose framework for the introduction processes of many different products. For example, the nose cowl PIP positions Rolls-Royce and its supplier in the 'Full Concept Definition' and 'Propulsion System Realisation' stages of the IPD process. The PIP selected for the case study is referred to as the 'Develop certified detailed nose cowl design' process. A top level IDEF0 [10] diagram of the nose-cowl is shown in Fig. 2. Being IDEF0 the diagram represents the process as an activity and shows flows of information and material both to and from the activity. Further, means by which the activity is carried out are shown as arrows entering the bottom of the box. Thus, using IDEF0 terminology, 'inputs are transformed into outputs by mechanisms under control'. For example, a 'CADDS file of validated nose cowl aero lines' is transformed into an 'Analysed and approved design of nose cowl' by a 'Design engineer' under 'Material requirements and constraints imposed on aspects of the nose cowl'. In addition, process characteristics, such as 'purpose of the process' (for example, 'to develop a product definition') and product related information such as a physical assembly of the nose cowl, can be identified together with relationships between them, such as, 'a meeting assigns a design engineer to develop a nose cowl physical definition'.

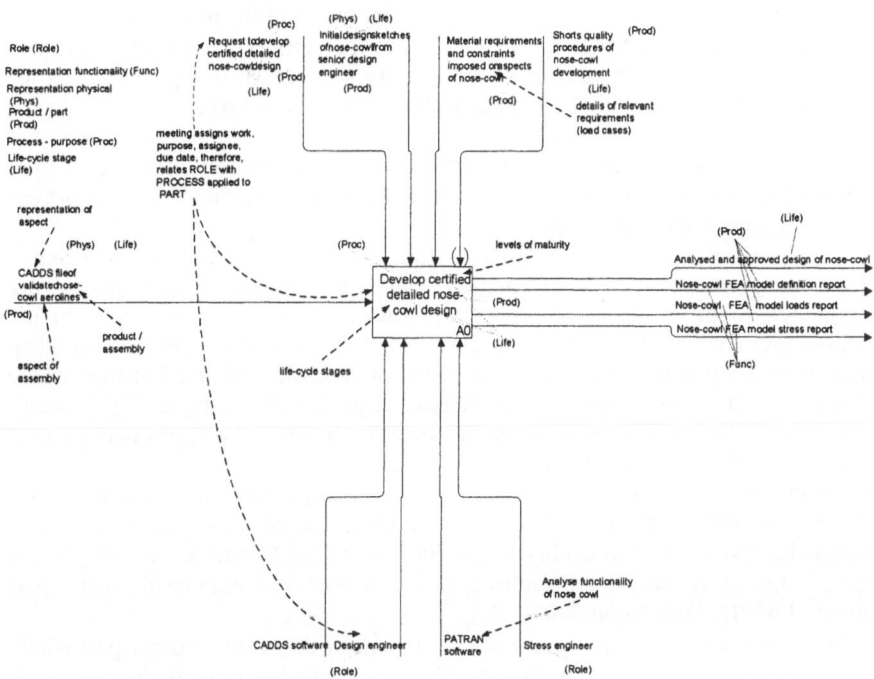

Fig. 2. Top level IDEF0 diagram of the nose-cowl design process

Process characteristics, information requirements, and relationships between information requirements, are overlaid onto activities and information flows in Fig. 2.

Key characteristics from the overlaid IDEF0 diagram are listed below.

- Role – a role performs an activity. This can be either a person, an application, or a machine, for example, software or an aerodynamicist.
- Artefact – an activity is performed on an artefact, that is, a part or an aspect of a part, for example, the design process can only take place if the activity has a part or concept of a part, on which to act. The part is the end product to be delivered to the customer, for example, nose-cowl. An aspect of a part is a sub-section of the part that is being acted upon by that activity, for example, the aerodynamic lines.
- Purpose – to carry out an activity it is necessary to know the purpose of performing the activity, for example, to develop the design of the nose-cowl.
- Result – processes 'result' in a product definition. This is the added value of carrying out an activity, for example, aerodynamic lines are the 'result' of performing the activity 'define aerodynamic lines'.
- Lifecycle – the processes support a phase in the product development lifecycle, therefore all activities may be attributed a context in which they are being carried out, for example, preliminary concept definition of the nose-cowl.
- Level of detail – provides further context to the activity description, as it provides a state which the output of the activity must reach for the activity to be completed, for example, 'Fully analyse the aerodynamic lines'.

It is these characteristics, the structure of connected activities, and their relationships to selected product structures that emerge from the definition process that is captured later in this paper.

Two product structures are used in this paper: a function structure and a physical structure. Other information, such as product requirements, are considered to be product data but not product structures because they do not exhibit the characteristics of a product structure that were outlined in [15]: namely, that a product structure is composed of elements and relationships where all elements are of one type and all relationships are of one type. Fig. 3 shows some of the relationships between elements of the functional and physical product structures, and a small number of product requirements.

Figure shows portions of functional and physical product structures and some of the relationships between them. For example, the physical product structure shows the "Nose cowl assembly Trent 700" as a part "Primary assembly Trent 700". In turn, this part is an assembly that has an assembly relationship with a part called "TAI 'D' Duct Trent 700".

The functional structure relates functions and in this case through part-whole [composition] relationships where the child is a sub-function of the parent. A functional structure is independent of the physical structure and can be created before the latter even exists. It aligns with Pahl and Beitz function structure [16]. The physical structure, on the other hand, relates definitions of physical parts. In this paper Bill Of Materials type information is used where two parts are related if one is composed of the other. Lastly, in this example, the heavy curved lines show relationships between the thermal anti-icing device and the elements of the physical structure that constitute this device.

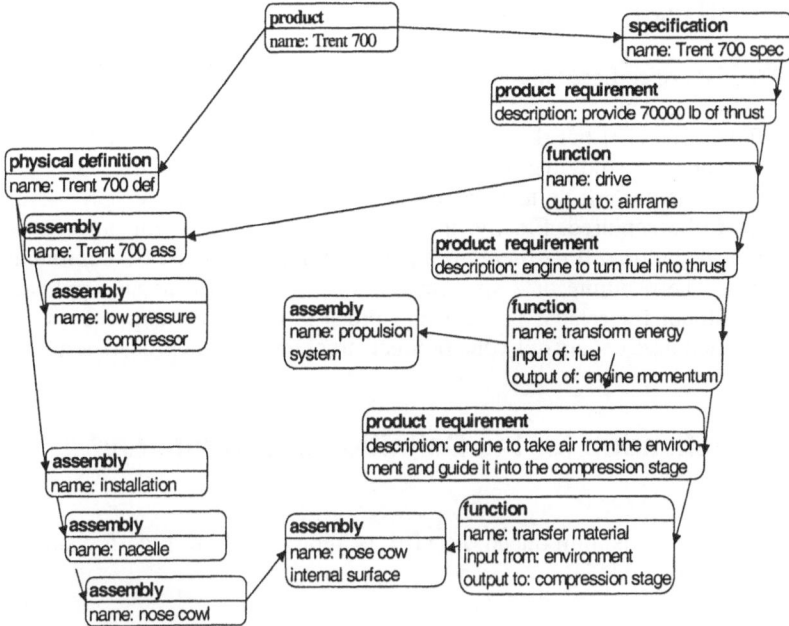

Fig. 3. Design rationale

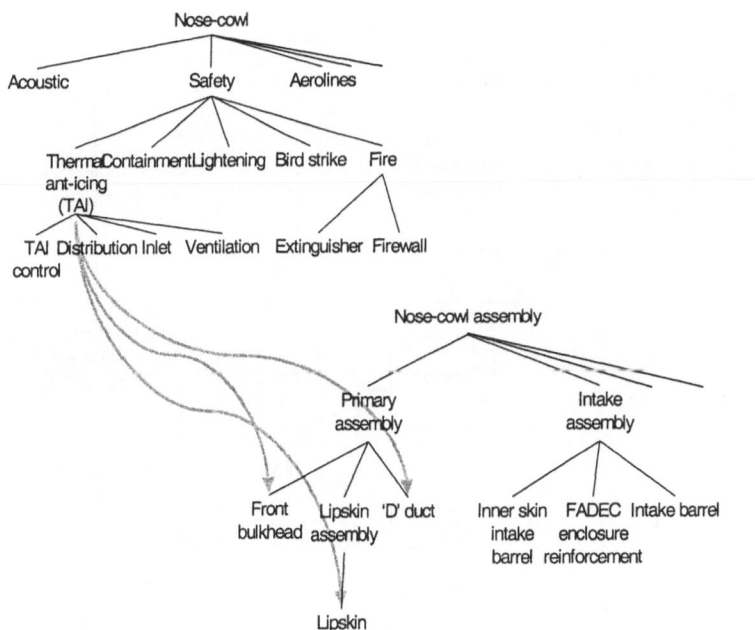

Fig. 4. Relationships between the product structures

4 Semantic Foundation

The STEP data specification architecture is used to frame the content of this section. Initially a general purpose data model, analogous to the STEP Integrated Generic Resources [17] but designed to relate product and process structures, is introduced. The case study described earlier in this paper is then used as a basis for information requirements to be satisfied if relationships between product and process are to be captured. Finally, the way in which the general purpose data model might be tailored [interpreted in the terminology of the STEP community] is illustrated. The resulting data specification can be regarded as an Application Interpreted Model for related product and process data. The scope of the data model presented here covers only one product structure and one process structure.

4.1
Data Specification for the Definition of Product and Process

A data model that is introduced in [18] is reproduced in Fig. 5 using the EXPRESS-G notation [12].

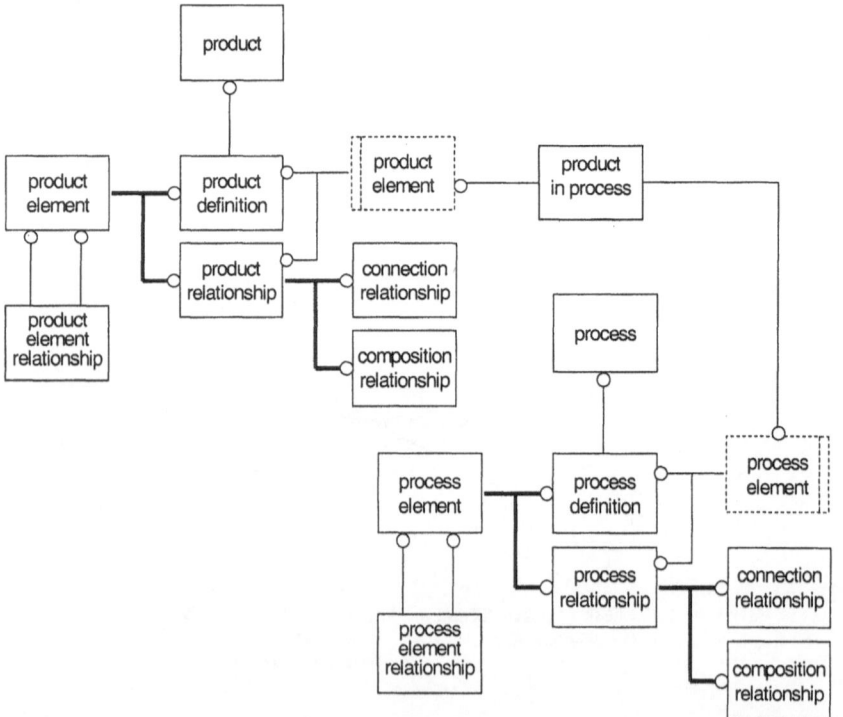

Fig.5. Data specification that can be used to capture both product and process

A single pattern is used for product and process structures. Readers who are familiar with the core concepts that underpin the normative data specifications of the STEP standard will notice parallels between the pattern and the STEP data specifications. A key difference is that in the STEP data specifications every relationship is existent dependent upon the two elements that it relates whereas in Fig. relationships are not. Thus, the data model in Fig. 5 treats both elements and relationships in a product structure as being of equal importance and allows relationships to be defined without the related concepts. This was initially introduced to support process definitions, for example as activities through IDEF0 diagrams, where flows [relationships] are often defined without the two (or more) related activities. A good example is any top level IDEF0 diagram such as the one given in Fig. 2. In product structures this additional functionality allows, for example, relationships between a product and its environment to be captured without the need to define the environment itself.

What the data model in Fig. 5 lacks is the contextual information that is needed for meaningful product and process structures. In STEP this is referred to as the application context, in IDEF0 such information is articulated through the purpose and viewpoint of the model.

4.2
Information Requirements from the Case Study

The product data context used in this small case study is based upon that of the STEP standard: hence the names of the product structures which relate to stages if the product life-cycle – a *product-definition-context* in ISO10303-41 [1].

- The process context is of more interest for this paper. Taking the case study given above the following may be regarded as elements of a "process definition context".
- Role – a role that performs an activity. This determines a viewpoint on the process.
- Purpose – to carry out an activity it is necessary to know the purpose of performing the activity.
- Lifecycle – a phase in the product development lifecycle – this may well be seen in the future as an aspect of the relationship between product and process.
- Level of detail – the extent of the process.

The remaining two key characteristics listed above, artefact and result, relate to the product (definition in this case) with which the process is concerned.

4.3
Information Support for the Case Study

The elements of the product structures can be captured using the *product-definition* entity from the data model in Fig. 5 whilst the arcs can be captured using the *product-relationship* entity. The activity box in the IDEF0 diagram can be captured using the *process-definition* entity from the data model in Fig. 5 and the IDEF0 flows can be captured using the *process-relationship* entity. Given these product and process structures the *product-in-process* entity can be used to establish relationships between the process and the product structure. A very small example instance, which includes property information in a format compatible with the STEP standard, is given in Fig. 6. An interesting point to note is that it is a representation of a property of a product definition that is related to the activity.

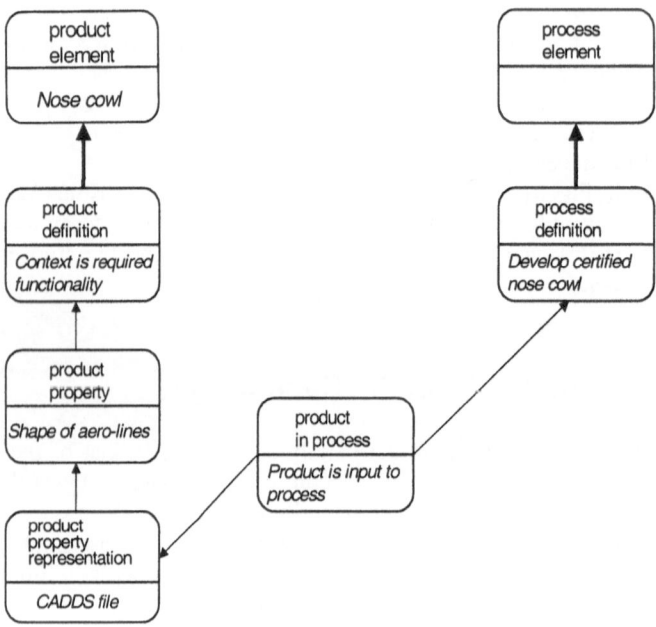

Fig. 6. Example instance of product and process

5 Concluding Remarks

This paper has reported early results of experiments intended to test the efficacy of applying product structuring know-how to the definition of processes. The resulting product and process models have a compatibility that allows them to be used as a basis for capturing relationships between the modelled product and process. Product and process structures captured using different underlying notations do not offer such an opportunity. For example, if a product were defined in terms of a STEP data specification then the product model would be an instance of an EXPRESS data specification; and if a process were captured using IDEF0 then the process model would be an IDEF0 model. Defining relationships between the two models would require a mechanism more abstract than either EXPRESS or IDEF0 and capable of capturing all of the concepts of each.

Relating product structures provides a possibility, in terms of an information system infrastructure, for achieving navigation through and across product structures, so that a change occurring in one product structure at a defined point in the process can be reflected in the other product structures. In practice there is a need to support multiple product structures and relationships between them. Work on relating product structures to support multiple process viewpoints is reported elsewhere [8]. Work on process structuring is at a far earlier stage than that on product structuring but the application of product structuring techniques to processes is yielding a better understanding of the different process viewpoints that might need to be supported in the future [19].

6 Acknowledgements

The authors wish to thank Rolls Royce for permitting the use of their case study in this paper. The ideas presented draw upon research that was carried out through two UK EPSRC/IMI projects: grant numbers GR/K96953 (CAPS) and GR/K78720 (SPEDE). Steven Trott's PhD research is supported by The Keyworth Institute of Manufacturing and Information Systems at the University of Leeds.

7 References

[1] ISO 10303-41, Industrial Automation Systems and Integration – Product Data Representation and Exchange – Integrated Generic Resources: Fundamentals of product description and support. December 1994.

[2] Andersson, K. A Design Process Model for Multiview Behaviour Simulations of Complex Products. Document DETC99/EIM0-9008 in Proceedings of the 1999 ASME Design Engineering Technical Conferences, 12-16 September, Las Vegas, Nevada, USA

[3] Bichlmaier, C., and Grunwald, S. PMM-Process Module Methodology for Integrated Design and Assembly Planning. Document DETC/DFM-8909 in Proceedings of the 1999 ASME Design Engineering Technical Conferences, 12-16 September, Las Vegas, Nevada, USA.

[4] Cohen, T., Peak, R. S., Fulton, R. E. Evaluating a Change Process Product Data Model for an Analysis Drive Supply Chain Case Study. Document DETC/DFM-9014 in Proceedings of the 1999 ASME Design Engineering Technical Conferences, 12-16 September, Las Vegas, Nevada, USA.

[5] Malmqvist, J., Schachinger, P. Towards an Implementation of the Chromosome Model – Focusing the Design Specification. International Conference on Engineering Design, ICED 97, 19-21 August, Tampere 1997.

[6] Andreasen, M. M., Hansen, C. T., Mortensen, N. H., On the identification of product structure laws. Proceedings of the 3rd WDK Workshop on Product Structuring, Delft University of Technology, Delft, The Netherlands, June 26-27, 1997. pp. 1-26.

[7] Svensson, D., Malmstrom, J., Pikosz, P., Malmqvist, J. A Framework for Modelling and Analysis of Engineering Management Systems. Document DETC/DFM-9006 in

Proceedings of the 1999 ASME Design Engineering Technical Conferences, 12-16 September, Las Vegas, Nevada, USA.

[8]   Trott, S. J., Baxter, J. E., McKay, A., de Pennington, A., Henson, B., Supporting Product Introduction Processes through Product Structures. Document DETC/DFM-8745 in Proceedings of the 1999 ASME Design Engineering Technical Conferences, 12-16 September, Las Vegas, Nevada, USA.

[9]   Peng, T-K, Trappey, A. J. C., Modelling Product Structures and Engineering Changes using EXPRESS Representation Scheme. In MED. Vol. 4, Manufacturing Science and Engineering, ASME 1996.
       ICAM Architecture, Part II, Vol. IV. Functional Modelling Manual (IDEF0). Report No. AFAWL-TR-81-4023. Available from Mantech Technology Transfer Center, WL/MTX Bldg 653, 2977, P. St. Ste. 6, Wright-Patterson AFB, OH 45433-7739, USA, June 1981.

[10] Information Integration for Concurrent Engineering (IICE) – IDEF3 Process Description Capture Method Report. Report No. AL-TR-1992-0057. Available from Mantech Technology Transfer Center, WL/MTX Bldg 653, 2977, P. St. Ste. 6, Wright-Patterson AFB, OH 45433-7739, USA, May 1992.

[11] ISO 10303-11, Industrial Automation Systems and Integration – Product Data Representation and Exchange – Description Methods: EXPRESS Language Reference Manual. December 1994.

[12] Westfechtel, W., Integrated Product and Process Management for Engineering Design Applications. Integrated Computer-Aided Engineering, Vol. 3, No. 1., 1996.

[13] Vaughan, C., (Editor), STEP/SC4 Industrial Data Framework. 22 October 1999.

[14] Ruffles, P. C., Project Derwent-A new approach to product definition and manufacture. Published by the American Institute of Aeronautics and Astronautics, 1995.

[15] McKay, A., A Framework for the Characterization of Product Structures. Proceedings of the 3rd WDK Workshop on Product Structuring, Delft University of Technology, Delft, The Netherlands, June 26-27, 1997. pp. 75-85.

[16] Pahl, G., Beitz, W., Engineering Design. Design Council, London, 1984.

[17] ISO/DIS 10303-1, Industrial Automation Systems and Integration – Product Data Representation and Exchange – Overview and Fundamental Principles. December 1994.

[18] McKay, A., de Pennington, A., Towards an integrated description of product, process and supply chain. Accepted for publication in International Journal on Technology Management.

[19] Baxter, J. E., Bloor, M. S., McKay, A., An application of STEP technology to support BPI. Proceedings of European Product Data Technology Days Conference, Paris, March 1998.

Part III

Metrics and Methods for Modularity and Configurability

Record from 3rd Group Work

Antti Pulkkinen

Tampere University of Technology
P.O.Box 589
33101 Tampere, Finland
e-mail: pulkkine@ruuvi.me.tut.fi

1 Introduction

The second day begun with a presentation of a method called Product Family
Master Plan by N.H. Mortensen from Technical University of Denmark. It was
followed by two presentations on matrix tools and a presentation on commonality
indices. After these presentations a third group work took place.

Because of the preceeding presentations, the groups discussed a lot of matrices
and commonality. Other methods, tools, and metrics were touched only lightly.
The following paper is divided into two sections. First, discussion on methods and
tools is described. Second, discussion on metrics is depicted. Finally some overall
remarks on the discussion is presented.

2 Records from Discussion on Methods and Tools

One group suggested step by step approach as a systematic modularisation
process. They argued that first a functional Design Structure Matrix (DSM^F)

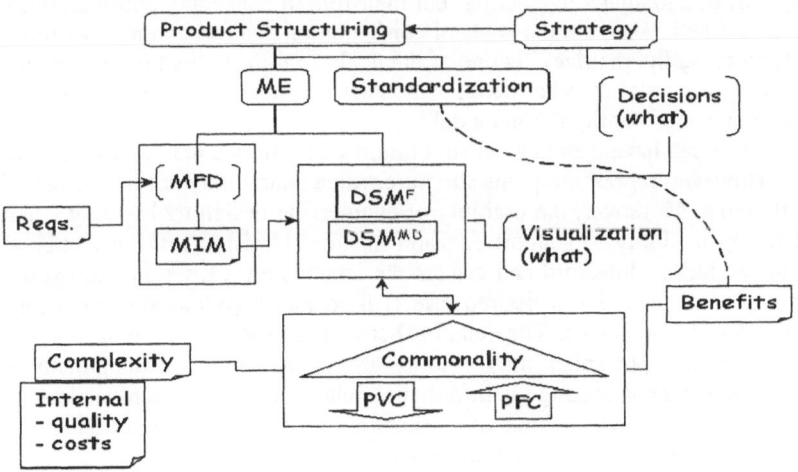

Fig. 1. Relation between methods, tools and metrics

should be developed. The second step would be deriving a similar matrix with condensed module drivers (DSM^{MD}). Thirdly, determining commonality indices should follow (see Fig. 1).

The selection of product structuring methods is a strategic decision in a company (see Fig. 1). Methods (like Modular Engineering or standardisation) are implemented by having certain procedures (e.g. Modular Function Deployment, developing Product Family Master Plan) and tools used in the procedures.

2.1.
Product Family Master Plan

The method of capturing configuration knowledge with Product Family Master Plan received quite a lot of attention. Participants stated that it supports configuring, which is based on both selections of functions and setting parameters for modules. It was seen as a language to be used for describing configuration knowledge. The case used in describing the method suggested quite astonishing results by proceeding from informal state to formal state in 10 days. However the case was not described in details and the suitability of the method to other approaches remained unanswered.

2.2
Matrix Tools

The discussion about the tools was an offset to presentations by Blackenfeldt and Järventausta. Thus, expressing product entities and their relations with matrices received a lot of attention in group work. Especially different forms of Design Structure Matrix were widely discussed.

Tools based on describing elements and their relations with matrices are certainly part of Modular Engineering, but their role in enhancing configurability is not well defined. What are the goals of configuration and can an engineer reach the goals by using the matrices? Is one of the goals a product structure, which can be used in configuration? What about a modular architecture that supports long term management of configuration model?

The product can have a strategical structure, which offers a new way to look at products. However, representing this structure with a matrix may not fit engineers' way of thinking. Moreover, the usability of engineering design tool is very much dependent on its ability to support decision making. Decision making in ME is related to product architecture and effects the architecture causes in life cycles. One very general goal (i.e. a desired effect) is to make savings (i.e. to reduce costs) in business processes. The relation between matrices and savings is not clear. Instead, the presented matrix tools are being used for documenting, communicating and condensing the information related to product architectures.

However, the rationale behind the structure is sometimes lost, because the reasons for making relations are not documented.

There is not a well-defined procedure on creating a matrix and many questions, like how to select the elements and to fill in the relations in DSM, remain unanswered. Also the criteria for deciding on whom should make the matrix, and what are the matrix entities and relations was not thoroughly communicated. Which are the product structuring or development phases, where the presented matrices can be used?

A static description of entities and their relations is the result of matrix tools. A hierarchical structure is supposed to be aligned with this description. However, a one structure may not be enough. In the discussion there were several opinions that suggested use of many (superimposed) structures. Yet, there are practical implications of the usability of matrix methods.

It was also sought after if the algorithm presented by Järventausta could be used on matrices by Blackenfelt? This is not possible. Also the term technical solution in the MIM-matrix was not clear. Is it a physical or a functional description of a product, or can it be both?

Business Process Re-engineering literature emphasises often the implementation of new, streamlined processes by introducing new software tools. However, relation between IT-tools and engineering processes is not very much studied in product structuring research. Moreover, the relation between product architectures and business processes can hardly be described with the presented tools.

3 Records from Discussion on Metrics

3.1
Criteria for Modularity and Configurability

It was stated that modularity is not a discrete property (i.e. no clear distinction between modular and integral system exists). Rather, it is a continuous property that could be measured with relational criteria. The criteria *modularity degree* was suggested for this purpose. However, none of the groups did attempt to define the criteria.

As product configurability relies partly on modularity, it is also a relational and continuous property. However, there are other factors contributing product configurability. These factors were not much discussed in groups. Instead, it was only suggested that configurability might be related to independency of customer selections in sales and ease of knowledge base (i.e. configuration model) maintenance.

Module drivers are abstractions about reasons for making certain product entities modules. The drivers provide generic views on issues like product strategy, life-cycle phases, re-use and sharing. The set of applicable module drivers is usually adapted to company specific situation. So far, any measuring method using module drivers has not been presented. However, module drivers cannot be directly linked with business objectives (e.g. lowering costs, increasing market share).

For making more comprehensive view on the modularisation reasons, it was suggested that research should concentrate on finding more of these company specific reasons and relating modularity effects to product life cycles. Another unclear issue is the role of module drivers in implementation of ME in a company.

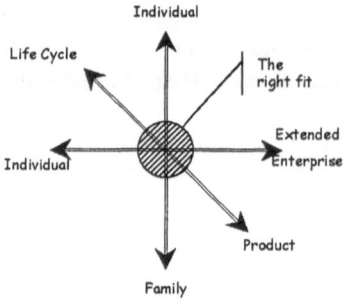

Fig.2. Design degree of freedom

Another suggested relational metric was the *design degree of freedom*, which is characterised by the right fit to certain factors (see Fig. 2). The factor were drafted as the degree related of product familiarity (individual vs. family), life-cycle (product vs. its' life-cycle) and organisation (individual vs. extended enterprise). The right fit was seen as an area demarcated by these factors.

3.2
Commonality

One group suggested that increased commonality is related to company's internal productivity (see Fig. 1) by reducing operational complexity. It was also stated that ME rationale is to move the complexity into more manageable area. This is supposed to lead towards decreased total engineering effort. Thus, it was suggested that the factors product complexity, degree of modularity, and total engineering effort are related, respectively.

Three aspects for commonality classification were suggested. First, it is a *relational property* between products, life cycles and organisations. Second, commonality means also properties and characteristics of products or variants are *re-used* by same organisation and life cycle. Third, commonality is describing *sharing* of the same product property when mirrored to organisation and life cycle. The question of the right amount of commonality seems to be generally irreconcilable, because it also involves business objectives.

As it is represented in Fig. 1, one group saw a relation between matrix tools (especially DSM) and commonality. They also suggested that having different kinds of commonalities yield to a number of benefits to company. So, by part standardisation, which is another form of product structuring leading to common parts, a company is diminishing its' possibilities for gaining benefits.

Commonality indices received also some critics. Some argued that any good commonality index have not been defined. Therefore, the purpose and use of indices is an unsettled matter. They also questioned commonality research effectiveness and focus.

One group contemplated that a design process is determining the physical solutions for given functionality. Thus, common functions can be harvested in structuring design processes (e.g. by re-using plans). With similar reasoning other commonalities contribute to the structuring of other processes. So, the context of a commonality determines its' contribution. However, commonality indexes do not have any context. Rather, it seems that in developing the indices a universal metric has been the mindset and the goal. It was suggested that a classification of commonalities is needed.

4 Concluding remarks

One method for capturing configuration knowledge and two matrix tools for developing modular product architecture were discussed. Modularity, configurability and especially commonality metrics were debated.

Impression from discussion is that the tools lack the procedure of use, i.e. the method they support is unclear. Thus, the rationale is many times lost and decisions cannot be traced back with the presented tools. The presented method may solve this problem, but more detailed case descriptions are needed. The current development of product structuring metrics does not seem to support very well design for configuration. Attempts on relating the metrics to context (e.g. life-cycle, business objectives) and to each other should be favored in future research. Some remarks for development of both the product structuring tools and the metrics were presented.

Modularisation by Relational Matrices – a Method for the Consideration of Strategic and Functional Aspects

Michael Blackenfelt

Royal Institute of Technology
Department of Machine Design
100 44 Stockholm, Sweden
e-mail: michael@damek.kht.se

Abstract. In order to get the most out of the potential benefits credited to product modularity, many aspects must be considered during the product structuring. This paper proposes a combination of two methods that have received considerable attention in the context of product modularisation, namely the Module Indication Matrix (MIM) and the component-based Design Structure Matrix (DSM), where the MIM focus on the product strategic aspects and the DSM on the product functional aspects. However, in order to integrate the methods efficiently, it is suggested that the MIM information is arranged into a DSM format, i.e. one result is the *strategic* DSM. The conventional component-based DSM is here referred to as the *functional* DSM. Moreover, the usage of the strategic DSM together with the functional DSM is exemplified with the modularisation of a vacuum cleaner. Finally, heuristics and metrics are employed to suggest a modular structure based on the information in the two multiple factor matrices.

1 Introduction

The last decade much research has been devoted to the area of product modularisation. Since modular products traditionally refer to products that fulfil various overall functions through the combination of distinct building blocks [1], product modularisation refers to the structuring of the product so that commonality and variety are well balanced. This in order to improve customer satisfaction by customisation, while retaining economy of scale. The research has provided a number of promising methods for product platform design, design for variety, etc. Today however, many researchers include incentives such as improved recyclability, organisation, supplier relations, product planning, etc., in the context of product modularisation. Symptomatically, the definitions of modular products have moved from the traditional definition above, to the more general definition; products that are composed of building blocks chosen by company specific reasons [2, 3].

Although, the handling of product variety is still most central, the latter definition imply that modularisation may be seen as a term for the rational structuring of products, i.e. an expression for the Integrated Product Development approach to the product structuring activity. Such an approach is important for projects, where business and technical objectives must be united or traded for each other to get the most of the potential benefits of modularity.

Many of the recently emerged methods are complimentary and thus it may be beneficially to use them together. This research have focused on the integration of two matrix methods that have been frequently referred in the design literature, i.e. the Module Indication Matrix (MIM), which is the central part of the Modular Function Deployment (MFD) [2, 3], and the Design Structure Matrix (DSM) [4]. The former of these two methods considers strategic aspects, whereas the other considers functional aspects of the product structuring activity. However, when clustering technical solutions or components into modules, obviously both strategic and functional aspects must be considered. Product structuring should be done by superimposing structural viewpoints, e.g. the functional view and the product assortment view [5]. The structure is defined as the sum of a set of elements and their relationships in the theory of technical systems [6] and since the DSM format describes the products elements and their relations this paper will advocate the DSM format for the structuring activity.

Thus, the purpose of this paper is to outline how strategic and functional aspects may be considered based on two available methods, i.e. the MIM and the DSM. The foundation is given in Section 2 by introducing the two methods and by reviewing previous approaches to combine the two methods. In Section 3, the approach of this paper is described. It is shown how the MIM information is transferred to a DSM format, i.e. the *strategic* DSM. Moreover, in order to aid the modularisation, heuristics and metrics are suggested. This is followed by an example, modularisation of vacuum cleaner, in Section 4. Finally, the paper is discussed and concluded in section 5.

2 The foundation

2.1
Strategic Aspects – Module Indication Matrix (MIM)

Modular Function Deployment (MFD) is a procedure for modular design, which has gained interest from both an academic and an industrial audience through various publications [2, 3, 7, 8] but also through the implementation by engineering consultants [9]. The MFD procedure is supported by a number of well-known methods, however the central and novel part is the so-called Module Indication Matrix (MIM). In the MIM the technical solutions are characterised by their *module drivers* that support the grouping of technical solutions into a modular concept. These module drivers may be seen as the reasons for a technical solution to form a module and are strongly related to the definition of modularisation, which accompanies the MFD method [2, 3]: decomposition of building blocks (modules) with specified interfaces, driven by company-specific reasons. The drivers are: *Carry-over, Technology push, Planned product change, Technical specification, Styling, Common unit, Process / Organisation, Separate testing, Supplier available, Service / Maintenance, Upgrading* and *Recycling*.

In the MIM, each technical solution is assessed in terms of its module drivers. This eventually results in a matrix of relations between the twelve module drivers and the technical solutions, figure 1a. The relations are normally given the weights of "9", "3" or "1" depending on the strength of the drivers. The weights are summarised vertically for each technical solution and the technical solutions with the highest scores will form the module candidates or initiators. The remaining technical solutions should be grouped to these initiators; preferably by grouping technical solutions with similar module driver profile.

In this paper the module drivers are seen as the strategic aspect of the modularisation issue. Then the term *strategic* refers to the business oriented plan for a technical solution or part, and is used to differentiate these aspects from the pure functional aspects. Furthermore, in the original MIM the term *technical solution* is normally used, however depending on design theory school or level of abstraction and detail, conceptions such as *function carrier (organ, principle) or component (part)* may of course be used instead.

2.2
Functional Aspects – Design Structure Matrix (DSM)

Design structure matrices are relational matrices that have been used to analyse product development processes and product development organisations, for example [10, 11]. However, here the focus is on the approach described in [4], where the product structure is analysed in terms of its components (elements) and the interactions (relations) between the elements. This is done by firstly identifying the interaction type and secondly by giving each interaction a score. The interaction types are *spatial, energy, information* and *materials*. The spatial interactions are given scores as follows: physical adjacency is necessary for functionality (+2), physical adjacency is beneficial, but not absolutely necessary for functionality (+1), physical adjacency does not affect functionality (0), physical adjacency causes negative effects but does not prevent functionality (-1) and physical adjacency must be prevented to achieve functionality (-2). For interactions of energy, information and material type, *physical adjacency* is replaced by *energy transfer, information exchange* and *material exchange* respectively. After all interactions have been quantified, the elements may be clustered into modules by sorting columns and rows. The idea is that the co-ordination complexity of the development effort may be reduced if the elements are clustered so that the interactions predominantly occur within modules, rather than between the modules [4, 12]. Figure 1b illustrates a fictive DSM together with a suggested clustering into three modules. Of course there will always be some relations outside the modules that represent the preferred module interactions. It should, however, be noted that in [4] the term chunk is used instead of module and these chunks are allowed to overlap each other. However, in this paper the technical solutions should exclusively be a part of one module, as in figure 1b.

In the original approach [4] components has been used as elements of the DSM, but other conceptions may of course be used. In this paper the term *technical solution* will be used. Commonly, the DSM type has been specified by its elements, for example component-based or activity-based DSM [13]. In this paper, however, there is a need to differentiate two matrices, which both are component-based (technical solution), and they are therefore differentiated by the type of relations. Thus, the matrix will be referred to as the *functional DSM*, despite the spatial relation is not really functional in the meaning of the traditional relations (M, E, I) of a function structure.

Module drivers	Technical solution	Technical solution 1	TS 2	TS 3	TS 4	TS 5	TS 6
Development	Carry-over	3				9	3
	Technology push		9				
	Planned product change			3	9		
Product variants	Technical specification	9	3				
	Styling					9	
Production	Common unit			9			9
	Process/organisation				9		
Quality	Separate testing		1				
Purchasing	Supplier available	3		9			
After sales	Service/maintenance			3			3
	Upgrading				3	3	1
	Recycling	1					
	Sum of drivers	16	13	24	21	21	16
	Module candidates			*	*	*	

a

	TS 1	TS 5	TS 3	TS 6	TS 2	TS 4
TS 1	S M					
	I E		2	-1	2	
TS 5		S M				
	2	I E	-1	2	2	
TS 3			S M	2		
	-1	-1	I E	2		-1
TS 6			2	S M		
		2	2	I E		
TS 2					S M	2
	2	2			I E	2
TS 4					2	S M
			-1		2	I E

b

Fig. 1a, b. A principal Module Indication Matrix **(a)** and a principle Design Structure Matrix **(b)**. In the DSM, rows and columns are sorted and bold lines indicate three modules.

2.3
Previous Approaches for Integration of Strategic and Functional Aspects

When clustering technical solutions into modules, obviously both strategic and functional aspects must be considered. Since both MIM and DSM have received considerable attention by the research community and they target different aspects of the issue, it seems promising to let the two methods complement each other. Other researchers have also identified this [14, 15, 16]. In [16] it is stated that the two methods should be used complementary, where an initial clustering is done by a DSM followed by an assessment by using MIM. In [14], the MIM information is introduced for each of the technical solutions in the diagonal of the DSM, but that approach does not show how the technical solutions are related to each other in terms of the module drivers. The information is merely pressed into one matrix, which reduces the comprehension. In [15], the fact that one side of the two matrices may be shared is used to add the DSM to the MIM, i.e. the original method is left untouched. Also in [17, 18] the combination of the two methods is discussed. Thus, all these researchers have foreseen benefits in using the methods in a complementary manner but have left the original methods more or less untouched.

Other research that strongly relates to this is [19, 20], where multiple factors are discussed in terms of DSM, especially the life-cycle factors (service, post life intent) and the spatial factor. The relational scores are however changed from the original DSM. Also along these lines, recently an ambitious approach to consider life-cycle factors was outlined in [21]. Although not matrix based, this type of reasoning may also be found in [22]. These authors are thus focusing on the last module drivers of the MFD-method. Related is also the approach to identify modules from function structures by employing both function and variety heuristics [23] as well as an approach to define product architecture [24].

2.4
Heuristics and Metrics

In [19, 20] heuristic algorithms are used to help the clustering of parts as well as metrics to evaluate the goodness of the product structure. The heuristics are based on a clustering algorithm [25] to identify the modules based on the information in the relational matrices. Their problem include multiple factors where they apply the algorithm on each factor at a time and then compare the resulting structures to find the intersections. The algorithm is used for the strongest relations since the algorithm cannot consider a strength difference. Moreover, in order to describe the modularity they have introduced a metric to describe the correspondence between structures driven by various factors, but also a Cluster Independence (CI) metric, which is the ratio between the number of relations inside the modules and the total number of relations.

3 Suggested approach

3.1
Reformatting the MIM to the strategic DSM

The approach, or method, outlined in this paper is entirely based on the two aforementioned methods. The MIM, with its module drivers, are addressing strategic aspects of interest to various stakeholders and in various product life phases. DSM on the other hand addresses the issue of modularisation from a functional viewpoint by representing the technical system in terms of elements and their relations. Thus, depending on the type of elements, the DSM potentially describes a function structure, a working structure or a construction structure, according to [1]. However, in this paper the term *technical solution* will be used and accordingly the DSM describes a working structure.

Here the DSM format is chosen since it gives a general view of the structure, described in terms of the elements and the relations. The idea is that the strategic aspects may be represented in the DSM format by having the same elements as in the functional DSM, but the relations is expressed as the degree of module driver sharing between two elements, as described below. The elements that share drivers should be clustered to modules, which are accomplished by a grouping of the current twelve drivers and a reformatting of the information. This reformatting provides the opportunity to superimpose the strategic and the functional aspects.

Some of the twelve module drivers consider similar aspects. The module drivers *carry-over, technology push and planned product change* address the issue of whether the solution is stable over time or not. The next three drivers, *technical specification, styling and common unit*, address the issue of commonality versus variety. The module drivers *process/organisation* and *supplier available* deals with the issue of where to locate the effort. These groups have been labelled *Carry Over, Commonality* and *Make or Buy* respectively. These groups include module drivers that address the same issue, where some drivers are contradictory whereas others are supportive of each other. Typically, *technology push* and *planned product change* support each other whereas *carry-over* contradicts the other two. It may be phrased that the drivers are on "opposite sides of the same coin" as illustrated in Fig. 2. The remaining module drivers share many aspects in terms of how the modularity is implemented and thus it is believed that this group of drivers mainly support each other. This group is here labelled *Life Cycle* due to its focus on typical product life cycle issues. These groups may be referred to as the *condensed module drivers*, a classification that is previously discussed in [26, 27].

These groupings are then used to transfer the MIM information to a relational format of DSM type. Firstly, the condensed module drivers may be analysed in a condensed MIM where the contradictions between drivers are represented by plus or minus, e.g. for *Carry Over* a minus will be introduced for *technology push* and *planned product change*, whereas *carry-over* stays positive. Furthermore, it may be noted that the module drivers with the score of "1" in the original MIM has been omitted in the condensed MIM, because they will anyhow, in practice, only contribute marginally to the decisions. Secondly, when this reduction of the module drivers is done, the MIM information may be transferred into a relational format. For example, in the condensed MIM of Fig. 3, the technical solutions TS1 and TS2 have *Carry Over*-scores of "3" and "-9" respectively. In the relational matrix this is transferred to "-1" in the CO-positions (symmetrical matrix). Thus, the transfer to a relational format is accomplished and the matrix will henceforth in this paper be referred to as the *strategic DSM*, in contrast to the conventional *functional DSM*. This DSM formatting of the MIM information was initially touched upon in [28] and further elaborated in [27].

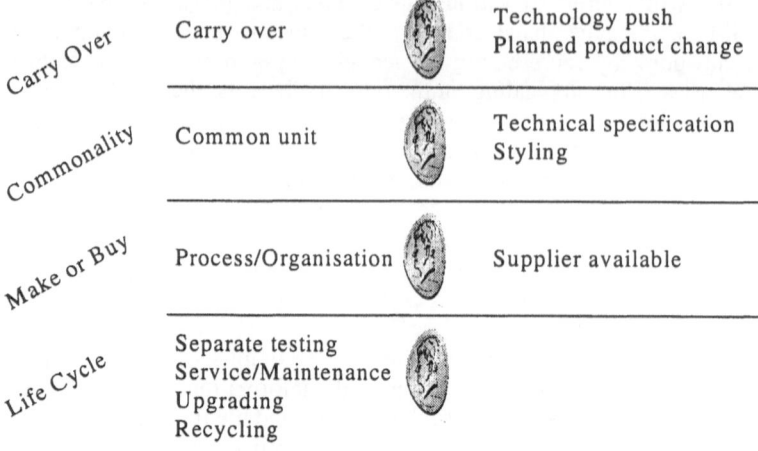

Fig. 2. The original twelve module drivers grouped into four "condensed module drivers".

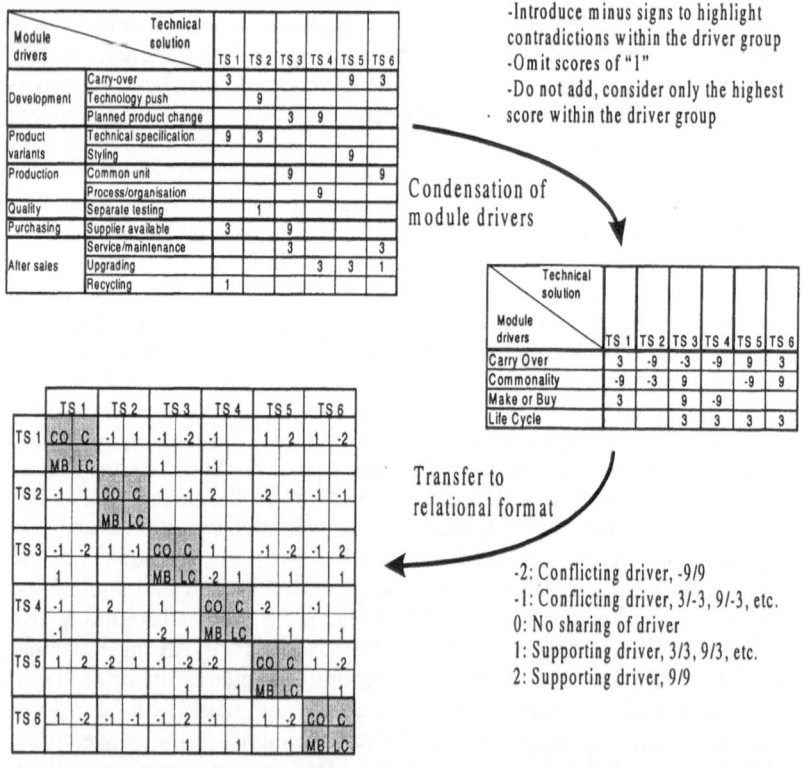

Fig. 3. The condensation of the MIM information into a DSM format.

The strategic and functional DSM may now beneficially be used together to analyse and create modular concepts. The information in the matrices gives advice on how to structure the products. Moreover, the two matrices together identify the strategic and functional trade-offs of a certain concept. If the identified trade-offs are not accepted, the structure, the technical solutions or the product strategies must be questioned and altered until a satisfactory solution has been achieved.

3.2
Heuristics and Metrics to aid the Module Clustering

Together the two matrices provide two *views* or *aspects*, each with four *factors*, on the structuring problem. Since the matrices contains much information heuristics and metrics would be useful for the structuring, as in [19]. However, in case the intersections between the optimal modules for each of the eight factors are searched for, only very small modules would be created, i.e. an approach that are less rigid are needed. Like [19], the heuristic approach in [25] is used to cluster technical solutions. However, here are various conditions applied on all factors or various sub-sets of factors before the actual clustering heuristic is applied:

1. Cluster technical solutions that *have at least one positive and no negative relation for both the strategic and functional matrices*. Similar condition is found in [20, 21].
2. Within the clusters from step 1, negative and positive strategic relations are separated by clustering the technical solutions having *only positive strategic relations*.
3. Finally, within the clusters from step 2, cluster the functionally coupled technical solutions, i.e. *technical solutions that have a material, energy or information relation, that is necessary (+2)*.

In case the modules still are considered to large, step 2 and 3 may be repeated at a more detailed level. In [17] a breakdown of the purely functional relations (M, I, E) are suggested, for example by differentiating between electrical energy and mechanical energy. Similarly the strategic relations may be represented at various levels as shown in Fig. 4. Thus, there may be reasons to divide a large cluster

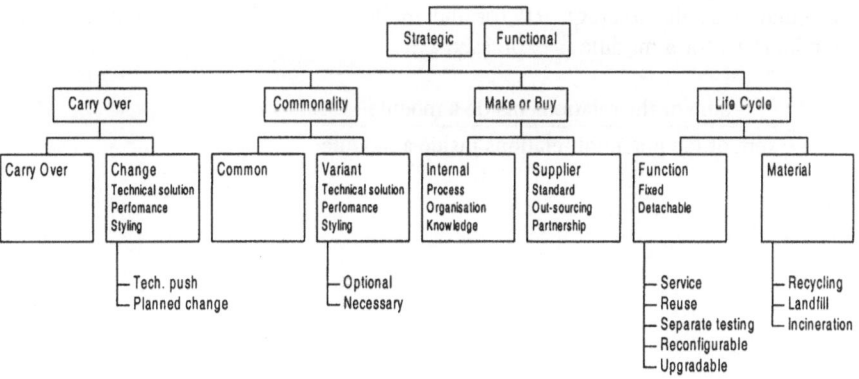

Fig. 4. A break down of the module drivers

characterised by for example future change in product plan modules and technology push modules.

In order to support the matrix structuring it is important with metrics, which reflect the goodness of the modularity. Metrics may be used to find the best modular structure but also to indicate improvements when changes are done to the technical concept or the product strategies. Metrics have been described by several authors however here the Cluster Independence metric [19] seems most interesting. It is adapted to include the values of the relations and is referred to as the Module Independence (MI) metric:

$$MI = \frac{\text{sum of the relations inside all modules}}{\text{sum of all relations}}$$

This metric is best used to study negative and positive relations separately, which means that the values should be maximised for positive relations and minimised for the negative relations. When there are no clustering, MI=0, and when all technical solutions are clustered to one large module, MI=1. This applies for both the negative and positive relations. In relevant and practical cases, however, the metric would be somewhere between 0 and 1. Moreover, the metric favours solutions where there are few negative relations inside the modules and few positive relations outside. The problem is to decide what is most important; low value for the MI_{neg} or high value for the MI_{pos}. The effect of negative and positive relations may be balanced by studying only one MI metric based on the sum of the relations. Such a metric would however favour the inclusion of a technical solution to a module as soon as the sum of positive relations to the other technical solutions in the module is larger than the sum of the negative relations. Moreover, when summarising negative and positive relations it should be ensured that the denominator is not zero or close to zero by using for example absolute values. Another and probably better approach is to study the relation between MI_{neg} and MI_{pos}, which is less forgiving to inclusion of negative relations into the modules.

A disadvantage with the MI metric is that it does not support further break down into modules within large blocks with only positive relations. Thus, here a complementing metric is suggested, the Average Ratio of Potential (ARP), which is calculated as the average RP for the modules of the product. The Ratio of Potential (RP) for a module is expressed as:

$$RP = \frac{\text{sum of the relations inside a module}}{\text{sum of the potential relations inside a module}}$$

4 Modularisation of a Vacuum Cleaner – an Example

4.1
The Reference Case

The use of the MIM has been illustrated in various publications by the vacuum cleaner case, for example [3, 8]. Now it seems logical to use this case to validate the strategic DSM and to illustrate the benefits achieved by using the strategic and the functional DSM together. However, according to the vacuum cleaner case and a few other published cases, sometimes the users of MIM have allowed a technical solution to have e.g. both *technical specification* and *common unit* as module drivers, although with different strength. The strategic DSM is based on a more stringent usage of the MIM, i.e. for each technical solution the choice must be done between the drivers positioned on opposite side of the same coin, according to the reasoning in the previous section. However, for the vacuum cleaner case it makes no practical difference for the validation of strategic DSM because the problematic scores are mainly "1"-scores which are not considered anyhow. The adapted MIM is shown in Fig. 5.

Module driver	Technical solution	Fan	Fan noise absorbent	Electric motor	Vibration damper	Motor noise absorber	Chassis	Bag	Filter	Thyristor	Thyristor button	Switch	Switch button	Cover	Cord with plug	Handle	Rear wheels	Front wheel	Bumper	Lid	Cord reel	Cord reel brake	Brake button	Bag lock
Development	Carry-over	9	9							9	3		3				3				9	3		
	Technology push						9	9																
	Planned product change																							
Product variants	Technical specification									1		1		3										
	Styling									9		9	9		9				9			9		
Production	Common unit	3	3	3	9	9	9	9	3	3		3					3	9	9		9	3		9
	Process / Organisation	9	9				9	9							9									
Quality	Separate testing		9						3	3														
Purchasing	Supplier available									9	9				9						9			
After sales	Service / Maintenance		3							9	3	1		1										
	Upgrading							9																
	Recycling		9			9									9							1		
	Sum of drivers	21	42	3	9	9	27	36	33	25	10	10	10	27	15	9	3	12	9	9	28	6	9	9
	Module candidates	*		*			*	*	*	*					*						*			

Fig 5. The MIM for the vacuum cleaner case, adapted from [3, 8]

In the MIM of the vacuum cleaner case only the main technical solutions are represented and not every single part which would be needed for the realisation of the technical solutions. Here 23 technical solutions are considered, but a vacuum cleaner of this type have roughly 70 parts. In [3, 7] it is argued that the optimum number of modules is 0.5-1 times the square root of the number of parts, depending on the assembly time of the parts to modules in relation to the assembly time of the modules to products; figures that should be considered as a rule of thumb rather than facts. Thus, since a vacuum cleaner of this type consists of around 70 parts it is suggested [3] that it should be divided in eight modules based on the module candidates chosen in Fig. 5: fan, electric motor, chassis, bag, filter, thyristor, housing and wire collector. In the conventional method, based on the module driver profile the remaining technical solutions are grouped to the module initiators, while keeping the technical possibilities and limitations in the back of the mind. The final module selection is shown in table 1 and the principal concept is shown in Fig. 6, [3, 8].

Table 1. Modules and the most decisive module drivers, [3, 8]

Module	Main module drivers
cover + handle + lid	styling
thyristor button + switch button + brake button	styling
chassis + bumper + rear wheels + front wheel + bag lock + vibration damper	carry over, common unit
Electric motor + thyristor + switch	separate testing, common unit
Cord reel + cord reel brake	supplier available
motor noise absorbent + fan noise absorbent	supplier available
fan	carry over
cord with plug	supplier available, considered as standardised part (no module)
bag	service
filter	upgrading, technology push

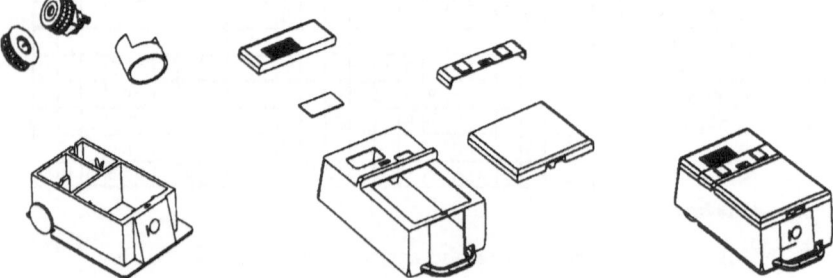

Fig. 6. Conceptual layout of the modularised vacuum cleaner [8]

4.2 Strategic and Functional DSM

From the information in the MIM, it is now possible to create a strategic DSM, Fig. 7. Furthermore, the modules of table 1 are represented in the strategic DSM by bold lines. It may be noted directly that most of the negative relations are outside the modules whereas a large proportion of the positive relations are inside the modules. Expressed by the Cluster Independence (CI) metric [19]; 13% of the positive relations but only 0.7% of the negative relations are inside the modules. However, from the matrix it may noted that more positive relations could easily be grouped into the modules without including more negative relations; this by enlarging the modules. In general the strategic matrix communicate the information provided in the previous section. It clearly shows one of the decided trade-offs of the vacuum cleaner case, i.e. the fact that the thyristor has been grouped with the motor and the switch despite the dissimilarity in the module driver *Make or Buy*. The thyristor has been "downgraded" from a module candidate to a standard procured part of the motor module. Moreover, the cover, handle and lid module could beneficially be grouped with the button module which implies that they are compatible sub-modules of the *top of the vacuum cleaner*.

	1	2	3	4	5	6	7	8	9	10	11	12	13	14	15	16	17	18	19	20	21	22	23
1 Cover	CC C MB LC	2	2	2	2	2	-2 2 2	-2	-2	-2	-1	-2	-2	-1	-1 2 2	-1 -2	-1 1	-1 2	-1	-2 -2	1 -2	-2 2 2	-1 -2 2
2 Handle	2	CC C MB LC	2	2	2	2	-2	-2	-2	-2	-1	-2	-2	-1	-1	-1	-1	-1	-1	-2	1	-2	-1
3 Lid	2	2	CC C MB LC	2	2	2	-2	-2	-2	-2	-1	-2	-2	-1	-1	-1	-1	-1	-1	-2	1	-2	-1
4 Thyristor button	2	2	2	CC C MB LC	2	2	-2	-2	-2	-2	-1	-2	-2	-1	-1	-1	-1	-1	-1	-2	1	-2	-1
5 Switch button	2	2	2	2	CC C MB LC	2	-2	-2	-2	-2	-1	-2	-2	-1	-1	-1	-1	-1	-1	-2	1	-2	-1
6 Brake button	2	2	2	2	2	CC C MB LC	-2	-2	-2	-2	-1	-2	-2	-1	-1	-1	-1	-1	-1	-2	1	-2	-1
7 Chassis	-2 2 2	-2	-2	-2	-2	-2	CC C MB LC	2	2	2	1	2	2	1	1 2 2	1 -2 1	1 1	1 2	1	2 -2	-1 -2	2 2 2	1 -2 2
8 Bumper	-2	-2	-2	-2	-2	-2	2	CC C MB LC	2	2	1	2	2	1	1	1	1	1	1	2	-1	2	1
9 Bag lock	-2	-2	-2	-2	-2	-2	2	2	CC C MB LC	2	1	2	2	1	1	1	1	1	1	2	-1	2	1
10 Vibration damper	-2	-2	-2	-2	-2	-2	2	2	2	CC C MB LC	1	2	2	1	1	1	1	1	1	2	-1	2	1
11 Rear wheels	-1	-1	-1	-1	-1	-1	1	1	1	1	CC C MB LC	1	1	1	1	1	1	1	1	1	-1	1	1
12 Front wheel	-2	-2	-2	-2	-2	-2	2	2	2	2	1	CC C MB LC	2	1	1 1	1 1	1	1 1	1 1	1 2	1 -1	-1 2	-1 1
13 Motor noise absorbent	-2	-2	-2	-2	-2	-2	2	2	2	2	1	2	CC C MB LC	1	1	1	1	1	1	2	-1	2	1
14 Fan noise absorbent	-1	-1	-1	-1	-1	-1	1	1	1	1	1	1	1	CC C MB LC	1	1	1	1	1	1	-1	1	1
15 Electric motor	-1 2 2	-1	-1	-1	-1	-1	1 2 2	1	1	1	1	1 1	1	1	CC C MB LC	2 1 -2 1	1 1 1	2 1 2	1 1	2 1 -2	1 -1 -2	-2 1 2 2	-2 1 -2 2
16 Thyristor	-1 -2	-1	-1	-1	-1	-1	1 -2 1	1	1	1	1	1 1	1	1	2 1 -2 1	CC C MB LC	1 1	2 1 -2	1 1	2 1 2	1 -1 2	-2 1 -2 1	-2 1 2 1
17 Switch	-1 1	-1	-1	-1	-1	-1	1 1	1	1	1	1	1	1	1	1 1 1	1 1	CC C MB LC	1 1	1	1	-1	-1 1 1	-1 1 1
18 Fan	-1 2	-1	-1	-1	-1	-1	1 2	1	1	1	1	1 1	1	1	2 1 2	2 1 -2	1 1	CC C MB LC	1 1	2 1 -2	1 -1 -2	-2 1 2	-2 1 -2
19 Cord reel brake	-1	-1	-1	-1	-1	-1	1	1	1	1	1	1 1	1	1	1 1	1 1	1	1 1	CC C MB LC	1 1	1 -1	-2 2	-2 1
20 Cord reel	-2 -2	-2	-2	-2	-2	-2	2 -2	2	2	2	1	1 2	2	1	2 1 -2	2 1 2	1	2 1 -2	1 1	CC C MB LC	1 -1 2	-2 2 -2	-2 1 2
21 Cord with plug	1 -2	1	1	1	1	1	-1 -2	-1	-1	-1	-1	1 -1	-1	-1	1 -1 -2	1 -1 2	-1	1 -1 -2	1 -1	1 -1 2	CC C MB LC	-1 -1 -2	-1 -1 2
22 Bag	-2 2 2	-2	-2	-2	-2	-2	2 2 2	2	2	2	1	-1 2	2	1	-2 1 2 2	-2 1 -2 1	-1 1 1	-2 1 2	-2 2	-2 2 -2	-1 -1 -2	CC C MB LC	2 1 -2 2
23 Filter	-1 -2 2	-1	-1	-1	-1	-1	1 -2 2	1	1	1	1	-1 1	1	1	-2 1 -2 2	-2 1 2 1	-1 1 1	-2 1 -2	-2 1	-2 1 2	-1 -1 2	2 1 -2 2	CC C MB LC

Fig. 7. Strategic DSM for the vacuum cleaner. Bold lines indicate the modules from table 1.

The modular concept above could of course not be created without knowing about the functional relations. In the original MFD method that information is considered as implicit. Here, it is argued that the functional DSM is a good carrier of the information, Fig. 8. In this case no negative relations are considered, not because there are non but rather because the technical concept is a conventional concept where the negative interactions are weak or easy to overcome, e.g. noise absorbents and vibration dampers are already included in the concept as *counteractions*, using the terminology in [17]. These counteractions are valued as "1" because they are not absolutely necessary for the functionality. As pointed out in [17], the spatial interactions are difficult to assess precisely without fixing the structure. Thus, here most of the spatial interactions are valued by a "1" to indicate that the relation is preferred but the technical solutions could be arranged differently. Moreover, a spatial relation has not been indicated between the technical solutions when any of the other three relational types have been indicated by "2", to avoid over estimation of the relations.

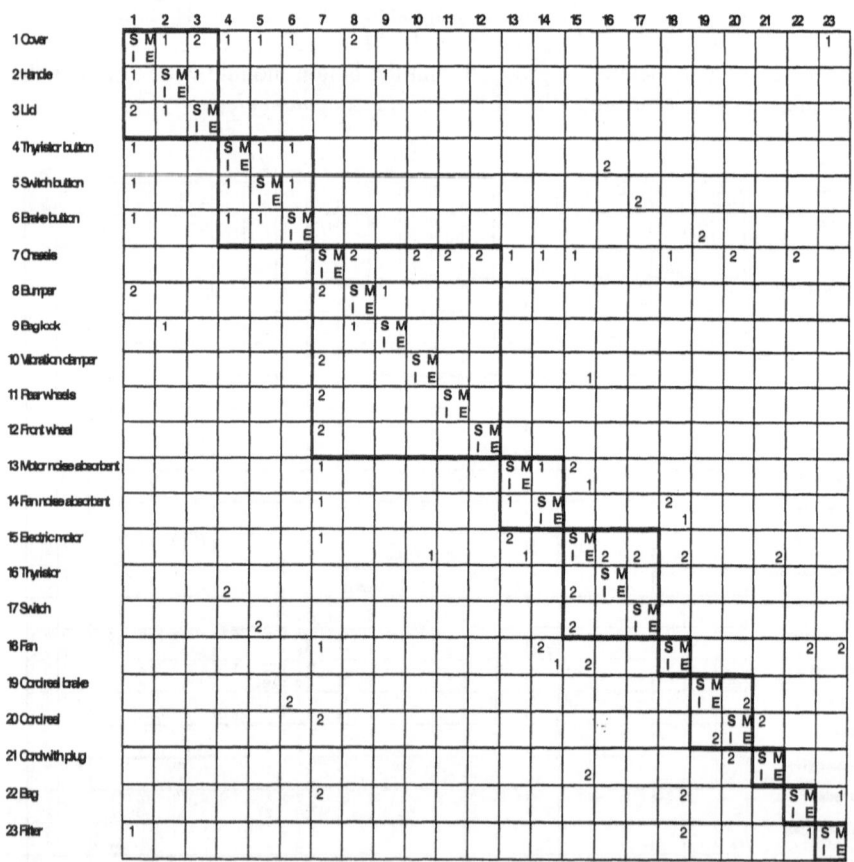

Fig. 8. Functional DSM for the vacuum cleaner. The modules from table 1 are indicated by bold lines

Generally, the functional DSM supports the clustering shown in Fig. 7; 1/3 of the positive relations are inside the modules. There will of course always be relations between modules. The two matrices together provide a tool to identify the trade-offs between strategic and functional requirements. If any of the negative relations inside or the positive relations outside the modules cannot be accepted, the technical concept, i.e. the structure, or the strategies must be questioned.

4.3
Heuristics and Metrics Applied on the Vacuum Cleaner Case

The combinations, that of the strategic and functional DSM provide a tool which helps to identify trade-offs and suggest changes when structuring the product into modules. The approach seems promising in itself however it would be even better in case algorithms could be applied to find a good modular structure based on the available information. Here, the three step approach described in Section 3 is tested.

1. The result achieved by clustering technical solutions which have at least one positive and no negative relation for both the strategic and functional matrices, is shown in Fig. 9. There are two clusters, *cover+handle+lid+buttons* and *the rest*, and three individual technical solutions, *thyristor*, *cord with plug* and *filter*. The three individual modules are reflected in the reference case, where the thyristor eventually is grouped with the electric motor.

2. For the remaining two clusters, negative and positive strategic relations are separated by clustering the technical solutions having only positive strategic relations, by once again using the heuristic algorithm. In the *cover +handle+lid+buttons* cluster there are only positive relations but in the other cluster the *bag* and the *cord reel* are separated out as individual modules because their negative relations to the other technical solutions in the cluster.

3. Now, for the remaining two clusters, the heuristic above may be employed to group the functionally coupled technical solutions, by clustering the technical solutions that have a material, energy or information relation, which is necessary (+2). Thus, here are the spatial relations excluded. In the *cover+handle +lid+buttons* cluster there are only spatial relations and thus it is untouched, however *motor+fan+switch* may be considered as a functional cluster within *the rest* cluster.

In the three steps individual technical solutions are separated out because the do not fulfil the conditions. However, among these technical solutions there may be suitable clusters and by running the three steps once again for these technical solutions, namely *thyristor, cord with plug, bag, filter, and cord reel brake,* the cluster *cord reel+brake* is identified. Thus the final result from the approach are the modules: *cover+handle+lid+buttons, chassis+bumper+baglock+damper +wheels+absorbents, motor+switch+fan, cord reel+brake;* and the individual modules: *thyristor, cord with plug, bag, filter.* The modules are similar to the modules of the reference case. In this case there are no reasons to further decompose, but step 2 and 3 may be repeated to separate technical solutions that have different type of strategic or functional relations as reasoned in Fig. 4.

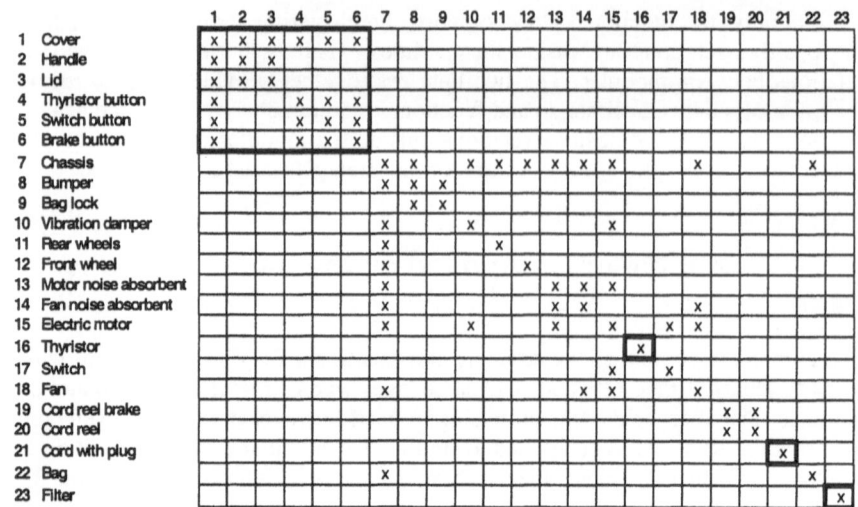

Fig. 9. Relations with at least one positive and no negative relation, considering all eight views. Bold lines marks potential module clusters from *the rest* which also forms a module cluster.

For the modules shown in Fig. 7 and Fig. 8 (reference case), $MI_{neg}=0.009$ and $MI_{pos}=0.21$. The metric favour solutions where there are few negative relations inside the modules and few positive relations outside the modules and thus the best combination for the vacuum cleaner case, when including all views, is probably the solution achieved after step 2 in the previous section, $MI_{neg}=0$ and $MI_{pos}=0.48$. That solution is based on two large clusters with only positive relations inside, i.e. by making smaller modules MI_{pos} will be degraded. It is actually possible to improve MI_{pos} even further by adding the individual modules to the clusters, but then on the other hand is also MI_{neg} increased. This may be balanced by studying the relation between MI_{neg} and MI_{pos}. Table 2 show MI values for four published cases where the MIM has been used; here is only strategic relations are included. In the meaning of the MI metric the modularisation of the vacuum cleaner is the most successful. In both the car ceiling case and the servo drive is MI_{pos} and MI_{neg} of the same order. In both cases technical realities may have prevented the clustering e.g. the technical solutions in the ceiling are placed far apart. In the drill motor case there is a low degree of clustering, which result in low values, but still MI_{pos} is much larger than MI_{neg}.

The ARP metric, in contrast to the MI metric, may be used to identify modules within purely positive blocks. The ARP metric has here been used on the sum of all relations from the eight views as shown in Fig. 10. Thus RP for the *cover+handle+lid* module, one of the modules marked with bold lines in Fig. 10, is calculated as (3+4+3)/3*16=0.208. Only half of the cells (3) are considered due to the symmetry and the other value in the denominator (16) reflects the potential value, i.e. maximum relation (2) times the number of views (8). ARP for all modules marked by bold lines is then the sum of the RP's divided by the number

Table 2. A survey of MI indices for some published MIM matrices; only strategic relations

Product	No. of tech. solutions	No. of modules	MI_{pos}	MI_{neg}	MI_{neg}/MI_{pos}	Source
Vacuum cleaner	23	7	0.18	0.009	0.05	[8]
Car ceiling	12	7	0.19	0.1	0.52	[8]
Servo drive	14	6	0.15	0.15	1.0	[8]
Drill motor (without optional modules)	21	15	0.06	0.006	0.1	[15]

of modules (6); ARP=0.192. The MI metric would favour modules marked by grey in Fig. 10, which correspond to the solution achieved after step 2 in the previous section. Due to combinatorics of the 23 technical solutions there are a vast amount of possible ways to create modules. Here, no intelligent algorithm have been applied to find the best ARP for the vacuum cleaner however the modules indicated by bold lines seems to be a good candidate. ARP separates out the buttons from *cover+handle+lid* as in the reference case, but on the other hand in difference to the reference case the individual modules *fan, cord with plug, bag* and *filter* are now clustered with other technical solutions. It should be noted that depending on whether all the vacuum cleaner is studied or only a subset, the division of modules based on ARP varies.

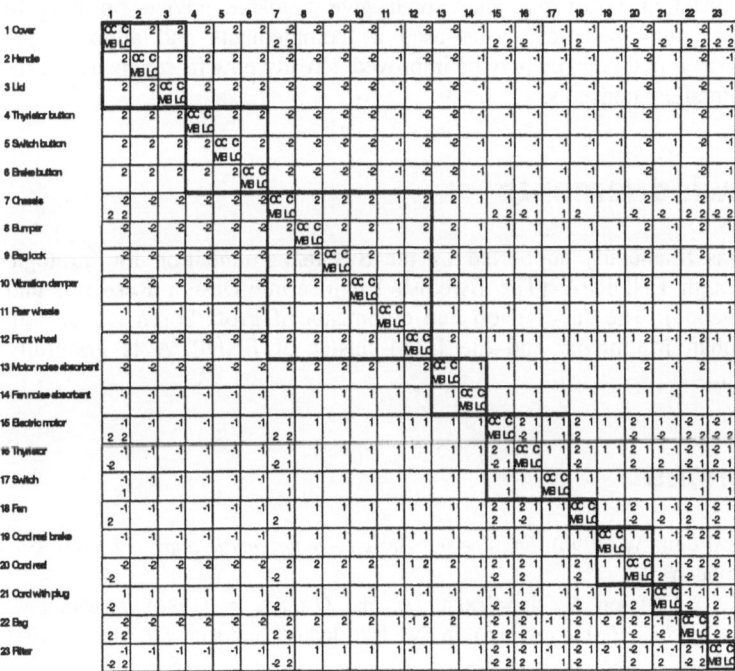

Fig. 10. The sum of all eight factors and potential modules based on ARP. The grey areas are the two potential modules identified based on MI

6 Discussion and Conclusions

The re-formulation of the module drivers and the introduction of a relational format to represent the module drivers improve the understanding of how the technical solutions are related. By introducing the strategic DSM and use it together with the functional DSM it is easier to identify trade-offs and potential improvements of the modular concept. The strategic DSM may be used as a complement to the MIM or as a separate approach. Moreover, based on the matrix information it possible to identify modules similar to the reference case. The discrepancies comes from the fact that the design team carry more information than what is available in the models, e.g. what relations are most important, what trade-offs may be done and which technical solutions are most important. However, the matrix based product models may be improved through an increased stringency. Currently the matrix models make no difference in importance between for example information and energy transfer or between mechanical energy transfer and electric energy transfer. The same is valid for the strategic relations. Of course it is possible to include weightings between for example functional and strategic relations, information and energy transfer, commonality and life-cycle; information which is needed for the final decisions on the product structure, but on the other hand it could be left to the designer to decide upon. Furthermore, here the qualitative heuristics and the quantitative metrics are used independently; future approaches may combine heuristics and metrics to optimise the structure. The two suggested metrics of the paper gives different result but since both give valuable advice on how to modularise they may beneficially be used as complements and although the approach described in this paper may be improved, already now its seems useful for generation of modular concepts.

7 Acknowledgements

This work was financially supported by the Swedish Foundation for Strategic Research through ENDREA (The Swedish Engineering Design Research and Education Agenda). The supervision and comments of Prof. Sören Andersson, Ass. Prof. Johan Malmqvist and the DTM-cluster of ENDREA is gratefully acknowledged.

8 References

[1]   Pahl G. and Beitz W (1996), *Engineering Design – a systematic approach*, Springer-Verlag, London

[2]   Erixon G., Erlandsson A., von Yxkull A. and Östgren B. (1994), *Modulindela produkten (Modularise the product)*, in Swedish, Industrilitteratur, Stockholm ISBN 91-7548-321-1

[3]   Erixon G. (1998), *Modular Function Deployment – A method for product modularization*, Dissertation, Royal Institute of Technology (KTH)

[4] Pimmler T. U. and Eppinger S. D. (1994), *Integration analysis of product decompositions*, ASME DETC, DE-Vol. 68, DTM, pp. 343-351

[5] Andreasen M. M., Hansen C. T. and Mortensen N. H. (1996), *The structuring of products and product programmes*, 2nd WDK workshop on Product Structuring, Delft, June 3-4

[6] Hubka V. and Eder W. E. (1988), *Theory of Technical Systems*, Springer-Verlag, Berlin

[7] Erixon G. and Östgren B., *Synthesis and evaluation tool for modular designs*, ICED, August 17-19, The Hague

[8] Erixon G., Fredriksson J., Romson L. and von Yxkull A. (1996), *Modulindelning i praktiken (Modularisation in practice)*, in Swedish, Industrilitteratur, Stockholm, ISBN 91-7548-460-9

[9] MMAB, Modular Management AB, (1999), www.modular-management.se, home page

[10] Steward D. V. (1981), *The design structure system: A method for managing the design of complex systems*, IEEE Trans. of eng. management, vol. EM-28, no. 3, Aug.

[11] Eppinger S. D., Whitney D. E., Smith R. P. and Gebala D. A., *A model-based method for organising tasks in product development*, Research in engineering design, no. 6, pp. 1-13

[12] Verho A. and Salminen V. (1993), *Systematic shortening of the product development cycle*, ICED, The Hague, Aug. 17-19, 1993

[13] DSM-MIT homepage (1999), http://web.mit.edu/dsm/index.html, December 1999

[14] Lanner P. and Malmqvist J. (1996), *An Approach Towards Considering Technical and Economic Aspects in Product Architecture Design*, 2nd WDK-workshop on Product Structuring, Delft, June 3-4

[15] Lange M. W. (1998), *CONCAD Bridging: Managing the integration of technical solutions in a modularised product concept*, 2nd International conference on integrated design and manufacturing in mechanical engineering, IDMME, UTC, Compiegne

[16] Jiao J. (1998), *Design for mass customisation by developing product family architecture*, Dissertation, Hong Kong University of Science and Technology

[17] Liedholm U. (1999), *Analysis of purposeful and incidental interactions*, ASME DETC99/DTM-8771, Sept. 12-15, Las Vegas

[18] Malmström, J. and Malmqvist, J. (1998), *Trade Off Analysis in Product Structures: A Case Study at Celsius Aerotech*, NordDesign, Stockholm, 26-28 Aug.

[19] Newcomb P. J., Bras B. and Rosen D. W. (1996), *Implications of modularity on product design for the life cycle*, ASME DETC, Aug. 18-22, Irvine

[20] Coulter S. L., McIntosh M. W., Bras B. and Rosen D. W. (1998), *Identification of limiting factors for improving design modularity*, DETC98/DTM-5659, Sept. 13-16, Atlanta

[21] Gu. P and Sosale S. (1999), *Product modularisation for life cycle engineering*, Robotics and computer integrated manufacturing, vol. 15, pp. 387-401

[22] Marks M. D., Eubanks C. F. and Ishii K. (1993), *Life-cycle clumping of product designs for ownership and retirement*, ASME DETC/DTM, Albuquerque, pp. 83-90

[23] Zamirowski E. J. and Otto K. N. (1999), *Identifying product portfolio architecture modularity using function and variety heuristics*, ASME DETC99/DTM-8760, Sept. 12-15, Las Vegas

[24] Allen K. R. and Carlson-Skalak S. (1998), *Defining product architecture during conceptual design*, ASME DETC98/DTM-5650, Sept. 13-16, Atlanta

[25] Kusiak A. and Chow W. S. (1987), *Efficient solving of the Group Technology Problem*, Journal of manufacturing systems, vol. 6, no. 2

[26] Stake R. (1999), *A hierarchical classification of the reasons for dividing products into modules – a theoretical analysis of the module drivers*, Licentiate thesis, Royal Institute of Technology (KTH), Stockholm

[27] Blackenfelt M. and Stake R. (1999), *Classification of module drivers to support product modularisation by relational reasoning*, CIRP design seminar, Enschede, March 22-24

[28] Blackenfelt M. and Andersson S. (1998), *Wheel-based mobile robot prototype designed with a modular approach – a case study*, NordDesign, Stockholm, 26-28 August

Enhancing Product Modularisation with Multiple Views of Decomposition and Clustering

Sami Järventausta, Antti Pulkkinen

Sandvik-Tamrock Oy.
P.O.Box 100
33101 Tampere, Finland
e-mails: sami.jarventausta@sandvik.com, pulkkine@ruuvi.me.tut.fi

Abstract. Conceptual modelling method for representing product data description is presented. It is compatible with the existing database technologies and product data management systems. The method enables modelling product data hierarchies from several points of view and linking them together. It is argued that by searching optimal configuration ranges from different points of view optimum and combining them a company wide pareto-optimum can be found and applied in modularisation. In the search of optimum a provided modelling method is to be used. A modularisation method that applies the presented modelling method and matrix clustering is suggested. Both the modelling and clustering methods are tested with a case study.

1   Introduction

The business objective is to achieve total efficiency in every process and to cover the valuable market area. This requires both the ability to cover wide range of market segments with smallest effort (i.e. minimal internal variety[1]) and the ability to make rapid process changes. The first requirement can be responded by configuring (or customising) the product to fit the individual needs of a customer in the sales phases. The second requirement demands agility from product development and production[2]. We see that both the product and process structure should be flexible.

Earlier product structuring research indicates that providing views is a one way to manage the complexity in the field [3, 4]. Our aim is to provide concept, which is compatible with existing system and capable to provide needed views.

Configuration can be enhanced with modular engineering [5]. Modules are building blocks that are assembled together to yield the needed configuration [6]. The resulting products are flexible in sales, but modularisation requires systematic and often time-consuming activities in product development. Our goal is to accelerate modularisation.

The key question in this paper is: How to generate modular structures fast and apply them efficiently in business processes? This demands structuring auto-mation in modularisation. Thus, we are interested on the formal and well-defined tasks of modularisation, which could be automated.

We present a concept of method, which supports fast production change by mirroring product modularisation to the product life-cycle stakeholders. The method is based on PDM systems and matrix clustering. It allows different modularisation practises and purposes to be used simultaneously. The method generates a new modular structure for the product related information after the changed of input data has been given.

The idea to apply clustering as a means for modularisation comes from Pimmler and Eppinger [7]. They have presented a method, which involves three steps: 1) decomposition of the system into elements, 2) documentation of the interactions between the elements, and 3) clustering the elements into architectural and team chunks. Our approach is similar but involves also separate views for different stakeholders.

A way for describing the product decomposition from many points of view is needed. The descriptions of the product are then clustered and the result is used in designing details and defining the configuration knowledge. The objective is a method, which can produce module structure from a description of a product. This can be used as by stand-alone, but also as a base for the final structure.

2 Product Model

Product model in this paper is understood in following way:
A complete product model is the total information about the product. In modularisation it is used to reduce the amount of information, which is needed in specific activities of the business process.

Reducing the amount of information is needed for streamlining the engineering processes. The information of the product model has to be organised in a logical way. Here we present some conceptualisations to be studied further. With streamlining the processes we mean that we are not trying to handle all of information related to the product at the same activity.

Fig. 1. The concept of a product model.

In figure 1. the whole sphere represents the total information about a product. Product related information is often handled in PDM *(product data management)* systems, but a PDM system is able to handle only data inside the cubic, which represents the structured data. The whole sphere contains also scattered and unstructured data (e.g. customers talking to each other, rumours and newspaper articles about the product). This amount of data is also too large to be managed in present PDM systems.

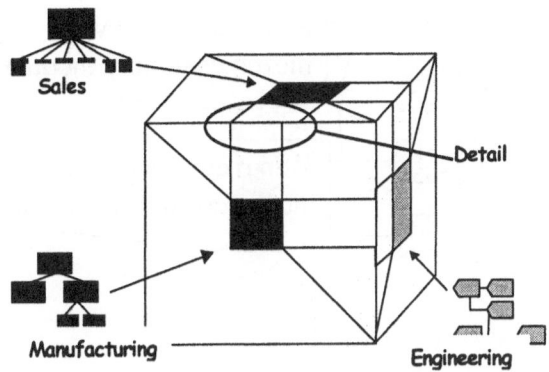

Fig. 2. Information presented in stakeholder views (adapted form [8]).

To illustrate different stakeholders' relation to the cube needs explanation. The faces of the cube stand for different point of views about the product. These points of views may be engineering design, manufacturing, marketing, logistic and etc. Of course, points can be found from every person related to the product. In order to reduce these points of views to manageable number, categorising persons to groups is needed. Mørup has presented the concept of external and internal stakeholders (marketing, engineering design and production). The cube is used because normally six major views should be enough. If one finds that he needs more views this method is expandable. In Fig. 2 we present three different flanks of the cube, each representing different perception of product [8].

Illustration in Fig. 2 presents the idea of organising data in a PDM system. In the centre of the cube there is an item, which is for example a diesel motor. This motor can be seen in different ways (depending on who is watching). The key to attach separate views together is in the edges of the cube. A concept how the attachment could be done is illustrated the Fig. 3.

The idea of concept presented in Fig. 3, is that single item can have different amount of relations to other items. In Fig. 4, we present how a single component is related to each view in the database with part-of relations.

The presented concept is based on very simple idea of using relations to establish different views. Also item related attributes can be used for this purpose (relation is also an item) and these attributes can be used for visualisations of relations. The visualisations are produced for different purposes. The method to handle views can be related to an existing PDM system. Because the present PDM systems are usually based on relational databases, the concept seems to be feasible.

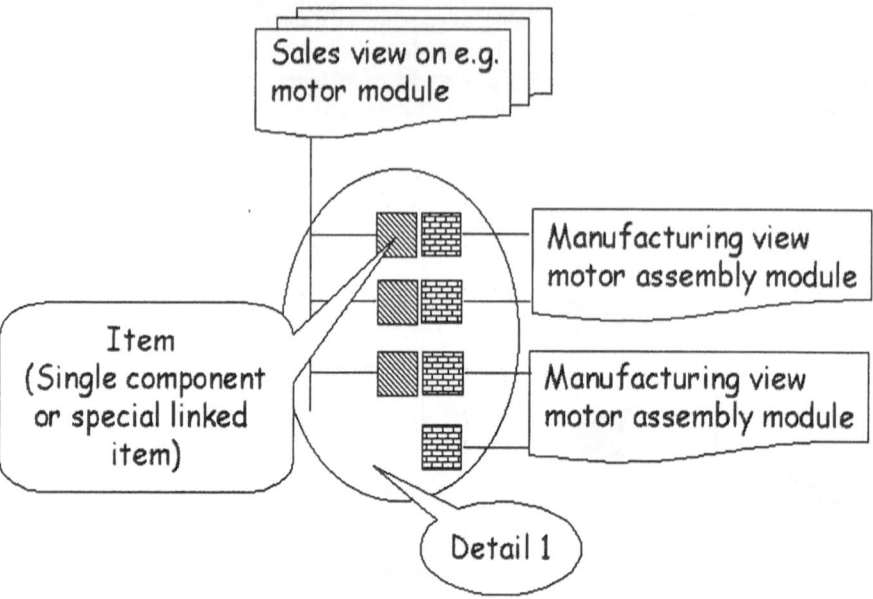

Fig. 3. Attaching different views.

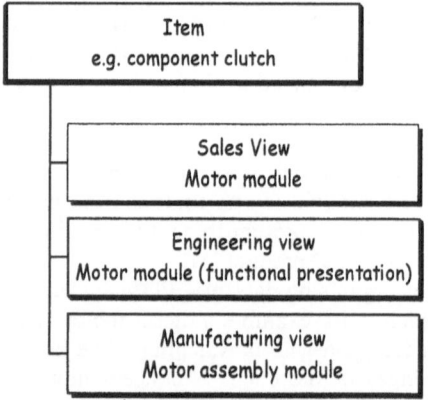

Fig. 4. Item-view relationship.

3 Configuration

A product that solely fulfils customer's demands is a one-of-a-kind product. A configuration cannot achieve this, but it usually is good enough and in many cases it is more inexpensive, delivered much faster, and more reliable. We may define configuring as follows:

Configuring is to vary the product in order to fulfil customer's requirements. Configuration process does not include novel engineering design tasks, but systematic variant design (that can be computerised).

Configuring product must enable a lot of combinations, which are in configuration range (see Fig. 5). The range has to be defined in advance. Product individuals, who are not found in the configuration range, need development of non-modules. Therefore they must be considered as non-configurable products. However, in practice business companies have to consider the products of projecting as a different form of configuration that is partially configurable products. In developing partially configurable products all the configuration knowledge is not formalised and captured.

Configuration knowledge is used in sales-delivery process, which can be characterised to be between projecting and mass production processes. Best process can be seen as a pareto-optimum, where order-delivery processes produce best profit for the company.

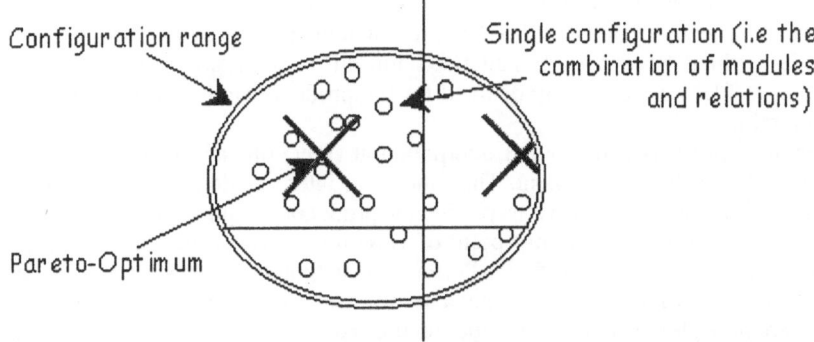

Fig. 5. The pareto-optimum in a certain product rance.

In Fig. 5 the idea of pareto-optimum is present in relation to configurable product. Single configuration stands for a specific product individual (i.e. implemented selection of modules and their relations). Pareto-optimum is a point where company's profit/sold product is at its' highest. However it is possible, that this configuration will not be sold to a customer.

The concept may be expanded to the different stakeholders, who can be included into consideration gradually. It is common, that first a company takes care of basic business issues (e.g. sales-delivery process) and after that it will consider other things (environmental, after-sales etc.). Often these issues are the interest of external stakeholders, such as government. Usually these issues were not known or considered in the beginning of technical development, when the emphasis was on developing the basic technical means like rock drilling technology. That is why pareto-optimum moves inside the configuration range (see Fig. 8) and configuration range should evolve according these changes along life cycle.

Configuration definition states that the creation of product variety is systematic. Modularisation can be seen as a general method to be applied in every product to support configuration. If product is not modularised, problems will occur with the management of relations in product model and rules in configuration model (i.e. the formalised configuration knowledge).

4   Modularisation

Modularisation is used to control and delimitate several attributes (e.g. functional properties and their counterparts in technical domain – characteristics) in one unit. This aids configuration process. We define the term module rather loosely:

Module can be attached to other modules. A complete configuration is a combination of modules.

The definition is not limited to any of the modularisation point of views. Modules, which are referred in this paper, may be established from sales, manufacturing, logistic or any other point of view. We presented the principle (see section 2) how modules, which have been modularised for different purposes, can be linked together in PDM system. Only demand for the product data is that it is presented in hierarchies.

Figure 6 presents a functional interpretation (functions and their relations) of existing rock-drilling equipment. The functional interpretation is main interest in configuration, when company is producing production equipments. From other stakeholder (e.g. manufacturing) point of view functions can be changed (e.g. to components and assembly orders). Different standpoints for modularisation lead usually into a different solution that has been attained with different methods. These are nowadays studied widely but no unified framework has been presented.

In Fig. 6 and 7 the functions are in many case very different resolution levels (i.e. levels of decomposition). This means that presenting the overall functional properties, which are needed in configuration, the resolution (e.g. in Fig. 7) is too detailed. To avoid the overload of information (i.e. observe a less detailed resolution level), we suggest reduction of information to encapsulated objects (i.e. modules). However, for further development the detailed resolution level is still needed. Therefore the configuration knowledge has to be gathered from product development and linked to configuration properties with part-of relations.

To fulfil module definition the views must have a connection, which can be established in PDM. If a product is very simple, the number of relations is small and the connection can be handled manually. Thus, with simple products a large and complex PDM system is superfluous.

However, in most cases companies have circulation of products and fast changes in personnel. This means that employer has to adapt to new tasks and be able to handle lots of information that is new to him. If this information is categorised, it is easier to handle the change. With the word agility we mean the ability to adjust into the continuous change. In modern business environment agility is necessity. It is the most important motivation for having a systematised and limited amount of information for each task. The paradox is usually an employee has to be an expert on very limited area, but he also has to be capable to communicate with other areas representing different viewpoints and resolution levels.

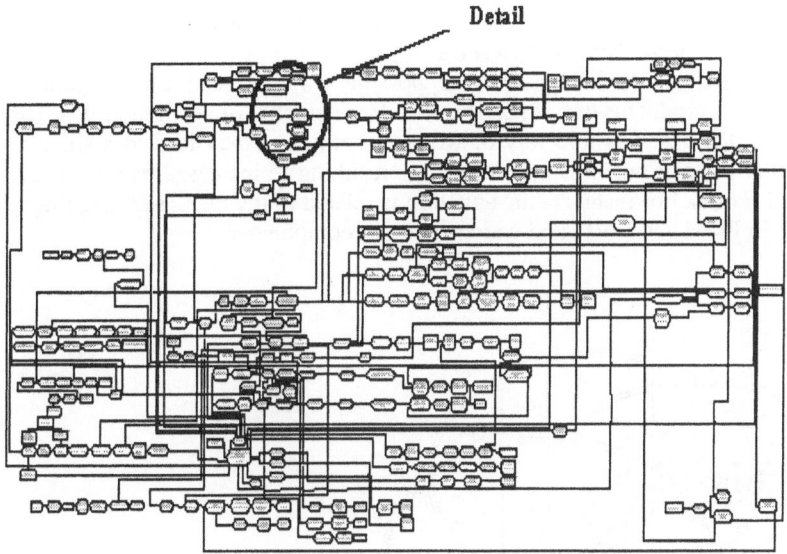

Fig. 6. Function chart of a rock drill rig.

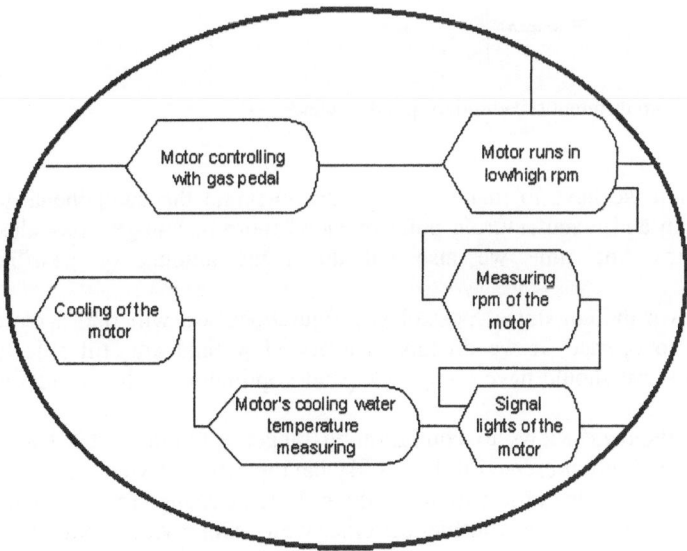

Fig. 7. Detail from the figure 6.

Modules are used in configuration, but the views that are not interesting for customer are simply hidden. However, these hidden views guide configuration in another way –providing values to attributes like price, delivery time, etc.

The presented concept depends on handling the relations between modules. On way to enhance the handling is to reduce relations into the minimum by making modules independent. This also makes it possible to create more variation (i.e. different configurations). With variation it is more likely that suitable configuration to customer need is obtained in sales.

Making modules as independent as possible can be seen as a separating process, where product is decomposed in to a large number of small modules. This alone is not the best possible way of working because it may lead to a situation where other parts of business are suffering – resulting to increased overheads and decreased profits. For instance, the situation is related to the increased handling of the items in PDM and MRP, and warehousing of components.

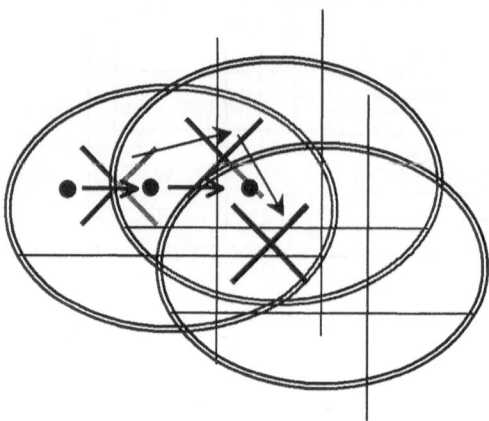

Fig. 8. Optimums from different stakeholders' point of views

After separation we have to make analysis and integrate the components to form (from design and manufacturing point of view) more meaningful modules. However, at the same time we also cut down the amount of possible configurations.

By cutting down the amount of possible configurations[1] we will find a more effective means to operate. If we do this in every view (and are still able to connect the views) we should have company's pareto-optimum for the product at hand.

Different stakeholders' views to configuration range, optimum and a single configuration inside it are presented in Fig. 8. By having different views on these issues we can have separated observations on them. When a change for a property is being made, we do not have to consider the other views. Other views' optimums

---

[1] This is usually the case in a company, which is approaching configuration from projecting.

stay as they are – only one view to optimum is altered. The pareto-optimum of the whole product is later changed according the change in one view. Result is that we have information about the change on local level, but also we have information on its affect on whole process

It is worthwhile to modularise the product in different points of view. The presented way emphasise any single point of view. Usually different views have different goals in modularisation. The goal of marketing department is usually the creation of variety, while engineering design and manufacturing are more interested in decreasing complexity and increasing kinship [5]. The pareto-optimum is a resultant of different views for the product.

5 Matrix Application

Modularisation rationale of this paper is to minimise relations between modules. Separating views enhances this, but it is not enough. For instance, having different views means that cabin sales attributes are silence, protection and comfortable, instead of muffle, shelter structure and padding[2]. Later we will integrate properties. In each view a collection of items into clusters should be done. The clusters of technical means (or collections) should match with customers' needs.

To a human being, drawing charts seems to be easy and natural way to support reasoning. This is implied by the fact that in many engineering tasks different kind of graphs is used – in electrical engineering, plant design, VLSI design, etc. However, in applying algorithms for clustering graphs seem to be cumbersome. Transforming graphs into another form of representation, which can be clustered by algorithmic means, seems to be the solution to this.

We may use chart for representing the cabin properties (boxes in Fig. 9) and the relations between properties (arrows in Fig. 9). In a matrix we always have rows and columns, which can be seen as factors. The factors have relations between each other. The relations are in a matrix marked in the cells (see Table 1).

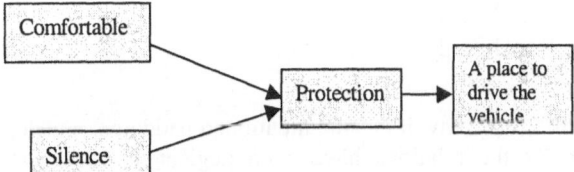

Fig. 9. A simplified chart representing the properties in cabin design

---

[2] In PDM the manufacturing view will be the second list as module-items (muffle, shelter structure, and padding).

Table 1. Chart transformed to a symmetric and an asymmetric matrix

A place to drive...	x			1	x			1
Comfortable		X		1	x			
Silence			x	1			X	
Protection	1	1	1	X	1	1	x	

It is possible to transform any chart to a matrix. Pimmler and Eppinger have used a symmetric matrix as a tool for searching modular structures. In their view a cluster represents wholeness, which should be the module [6]. To represent information about the course of relation (i.e. the direction of an arrow in a chart) an *asymmetric matrix* can be used (see the right side of the Table 1).

In the right part of the Table 1, the information about relation's direction tells for example that "A place to drive..." is defined by protection. If we want to express what we mean with "Protection" we may to define it is "Silence" and "Comfortable" – the values for "Protection". Usually these kinds of matrices are called *Design Structure Matrices* (DSM)[3].

Modularisation of product can be very different according to the different viewpoint (i.e. the shareholder and the resolution level of the product-related data). Next we will present a possible structuring from manufacturing point of view for the cabin-related items.

Table 2. Symmetric matrix for cabin (from manufacturing point of view)

Cabin	x	1	1	1
Padding	1	x		
Shelter structure	1		x	
Muffle	1			x

In Fig. 10 the chart looks different. Now it has the mentioned paddings, muffle, etc. Note that the directions of the relations have been neglected. Thus, the corresponding matrix (in Table 2) is symmetric. The clustering of the cabin has been done according to manufacturing point of view with rather coarse resolution. The rationale in developing the items and the relations is to collect assemblies, which are needed at the same time and in same place.

We see from table 2 that the optimum (from manufacturing point of view) is now strongest weighted from module Shelter structure/Protection. For configuring

---

[3] For more information see: http://www.mit.edu/afs/athena/org/d/dsm/.

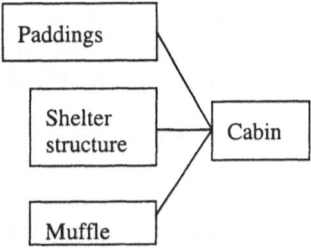

Fig. 10. A simplified chart representing the structure of cabin design

this means that when protection properties are selected lots of cabin's manufacturing characteristics are being selected. By defining a protection property we are affecting widely on whole configuration. This can mean that customer will not have an optimal solution for his/her needs.

6 Matrix Clustering

Matrix clustering means that the decomposed properties of a view[4] are now collected to groups. This can be done manually, but using algorithm is easier with fine resolution level or large number of details – in matrix columns and rows. On the other hand, using algorithm means that the user cannot interfere on clustering

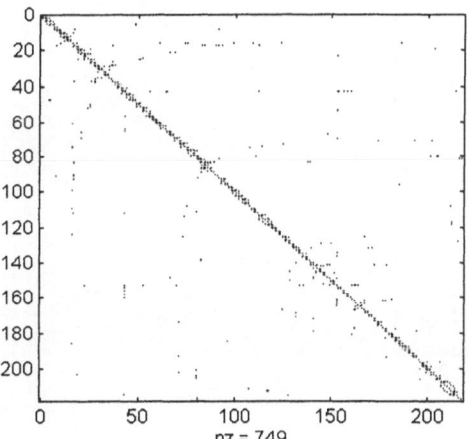

Fig. 11. Base data matrix

---

[4] This means also other attributes than those related only to a customer specific configuration.

results. In this paper we are using method, which from users' point of view is mechanical (i.e. an algorithm). Our interest is in how long we can use one algorithm and when we have to change to another algorithm or possibly to manual method.

The base data matrix (in Fig. 11) presents a transformation of the functional chart (from Fig. 6) of whole drilling rig. We have estimated that engineers deal with 240 different functions in configuring the rig. Lots of relations between these functions emerge. Here the relations have no direction and we use a symmetric matrix. Our goal is to find the functions, which have no or only few relations with other functions. Clustering begins with arranging the matrix in order to have relations as close the diagonal as possible.

Base data matrix is now arranged and result (Fig. 12) shows, which functions are the most next to the diagonal. The Cluthill-McGee algorithm is used for arranging. At this stage we have not integrated any of the functions.

The next step is to reduce relations and functions in order to have a limited collection of functions for configuration. In doing this, we neglect all of those functions that have only one connection to other functions. Typically these are additional properties (e.g. outtake pistol for compressed air). Naturally, this option has a relation to the pipeline of pneumatic system. However, with a closer examination it is seen this relation is only trough air outtake line.

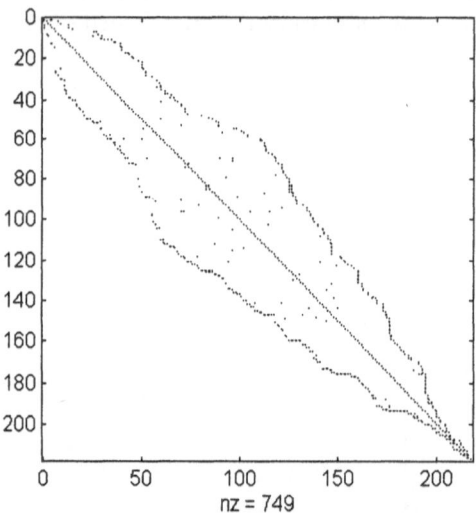

Fig.12. Arranged matrix

Pistol is now integrated in to the pipeline, because pistol's relations are only to the pipeline. How we define this action in sense of mathematics? Next illustration presents a mechanised method to find these kinds of groups from the arranged base data.

If this method can be described in terms of mathematics, computer can make the clustering by itself and demand of automation is fulfilled. Let us take a closer look on the method

The dark arrows beside the matrix present cluster, which is easy to find. X1 has one relation and it is with X2. Because matrix relations are like data base relations they are double directional. X1 can be compared to air outtake function or pistol component presented earlier. Cluster containing X1 and X2 is named after X2. In next round clustering is done so that the new X2 is clustered inside X3. This kind of cluster is named here as a 1. Degree cluster. These are found first and in this case 14 clusters are handled. Defining for these kind of clusters:

1. Degree cluster: Only ONE item holds relations outside the cluster

In case of X3 there is the only item inside the cluster, which holds relations outside the cluster. This is true, because X1's relations are with X2 and X2's only relations are with X3. Cluster has limited amount of relations (2) and they all can be found from one item. To find these kinds of clusters fulfils the demand of independent module, which has as few relations with other modules as possible.

However, clustering is not finished yet. We can find from the matrix about 20 1.Degree clusters. This means that chart's amount of items has reduced to 70 %. Thus, the rest of the items cannot be clustered with this method. To expand the method a second-degree cluster is defined:

2. Degree cluster: Only TWO items holds relations outside the cluster

This definition is presented also in picture 13: X4 and X5 are clustered inside X3 and the new item is X3. This is because X3 originally holds more relations than X4 and X5 hold together. Then X3 has more importance than X4 or X5, when determining the relations to other modules. This cluster holds then information about relations outside the module from X3 and X5 and therefore fulfils 2. Degrees cluster definition. X4's relations are not anymore needed, because X3 already holds relation to the same item than X4. So, clustering of X4 inside X3 is not increasing amount of relations for the cluster. If needed 3. Degree cluster can be defined same way as before, but for the chart (Fig. 6) it was unnecessary.

Now we have modules, which are interesting in configuration point of view. Customer should only be interested about the properties like air outtake. How we manufacture this property is not observed in this chart, but in another. Anyhow, we must have a connection between these two charts. This is the key for mastering different views together. Remembering cube, which were presented earlier (Fig. 2), we have different views and we need then connection between views.

Evaluation of the Clustering

When clustering is done every item is clustered with some other item. Final result is one item, which has collected all other items to it. The structure is hierarchical and therefore present PDM systems can handle it. Algorithm has no opinions about optimal modules, if we have not defined those things inside it. Probably the result is "raw" and needs some fine-tuning in order to be useful. To check the result of clustering we study what kind of modules it suggests.

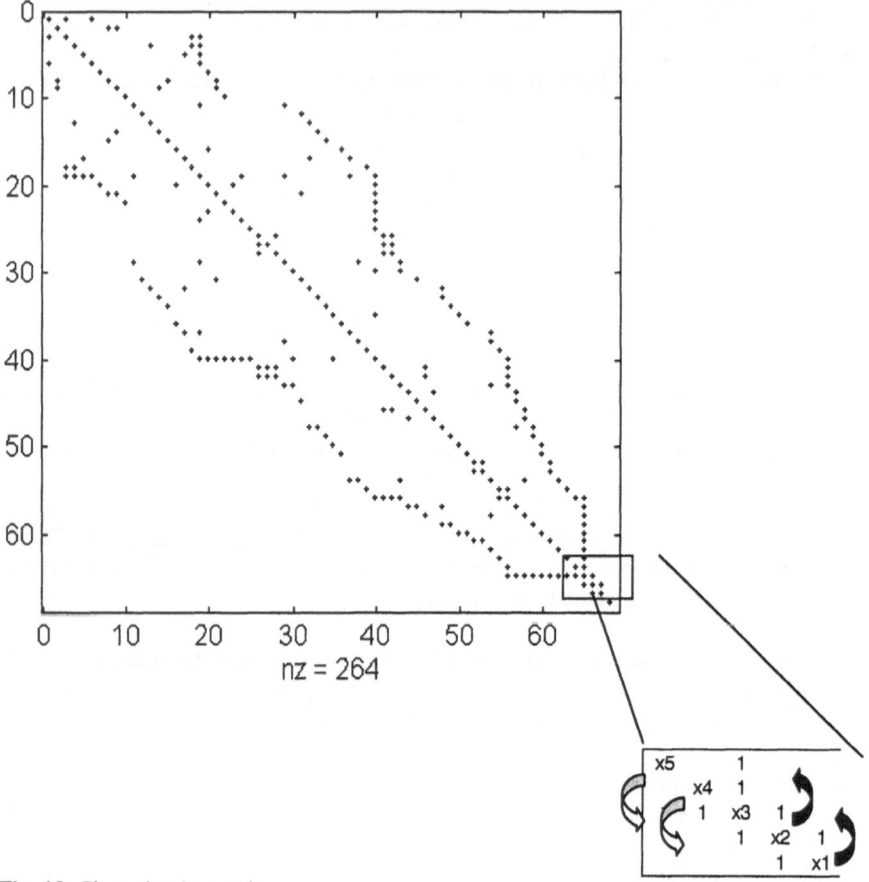

Fig. 13. Clustering in matrix

The result consists from 30 modules. These modules are all clustered inside one item, which is main electric box – the main electric box has relations more than any other item. This is easy to understand because drilling rig uses electricity as an energy source for almost everything. On the first level we have one item "Electric main box" and on the second level we have 29 items. In order to have useful product structure we have to have on the first level item "product". This means that Electric main box have to be divided inside all other clusters. This is tearing[5] process repairs cluster, which are too big ones. Every item inside the item *electric main box* has to be repositioned in to other clusters. E.g. electric control of main powerpack has to be clustered with powerpack itself.

Second manually handled issue is "false" clustered items. This means that some cases secondary function (generally chart's item) clusters inside itself primary

---

[5] Term *tearing* is presented in DSM methodology and it has a purpose to cut relations inside iteration cycles.

vital functions. This is because method is investigating only the amount of relations and no quality of the relations. This problem may be avoided partly by gathering into the matrix only the items, which have relation "junction". For example in Fig. 7 item "Motor's cooling water temperature measuring" is not included in matrix, because it has only one input and output relation. "Junction" item is for example "Signal lights of the motor", because it has two inside relations. Reducing items eases clustering and also gives more value for items, which have relations to many other items. Despite of this some cases were found were the importance of secondary function had been increased by accident — for example wiring had clustered inside itself more primary functions, such as water pump. These kinds of errors are due to the lack of information given for the method. If algorithm is used, it should have information about the quality of relations and the ability to cluster according to them.

Despite these manually handled issues, the main task was achieved. Modularisation of the product related information was structured easily and almost mechanically. Method did (excluding "false" clustered items) not depend on the employee. Everybody can and will have the same "raw" result and only difference is fine-tuning of structure. Method is still quite slow, because it takes two to three days to cluster whole matrix manually (45 clustering rounds) with aid of computer. The used applications were Matlab and Excel (in which the clustering was done manually). First one clustering was done, and then it was rearranged with Matlab by Cluthill-McGee algorithm. This algorithm eased manual clustering by ordering items as near diagonal as possible (Fig. 12 and 13). After clustering arranging another round of manual clustering was done.

With the help of algorithm the manual clustering is easier to do – relations are nearby each other. In order to have this method useful; also clustering should be computerised. In that way the base data matrix could be transferred immediately and the results could be seen be seen after the whole clustering process. Then the process would need people to do the clustering and the spent hours would be reduced.

7 Conclusions and Further Studies

We have presented a concept how to describe product model in different views and levels. In this paper a product model is a combination of product related information and the viewpoint to product model is usually a reduced amount of product related information. The reduction of information can be made with determining needed resolution level and viewpoint for each product development, manufacturing and sales task. We argue when modularisation is used in every view and modules are used in business processes, business should operate in pareto-optimum

The presented method can produce a module structure from a description of a product.

We have been able to generate a module structure very fast for the configuration process. Defining the needed information in sales-delivery process based on configuring and classifying the stakeholders does this. The presented concept is based on the viewing abilities of PDM systems.

The presented method is based on clustering operation and it allows applying different modularisation purposes at the same time. Every view can have different kind of information and also different kind of method to modularisation (possibly without matrix clustering at all). However clustering is a method to "automate" modularisation and make it efficient for implementing changes.

We suggest that different algorithm for different modularisation purposes are to be studied– e.g. one algorithm for functional description and for manufacturing description. Also the possibilities to expand the method to cover external stakeholders, like environment issues, should be studied.

8 Acknowledgement

The authors want to express their gratitude to the financial support from Technology Development Centre of Finland (TEKES), Tampere University of Technology and Technical University of Denmark.

9 References

[1]   Elgård, P., Miller, T.D., Designing Product Families. In: *Design for Integration in Manufacturing*. Proceedings of the 13th IPS Research Seminar, Fuglsoe 1998. ISBN 87-89867-60-2. Aalborg University 1998.

[2]   Boynton, A.C., Victor, B., Pine II, B.J. *New competitive strategies: challenges to organizations and information technology*. In IBM Systems Journal, Vol 32, N 1, 1993. pp. 40-64

[3]   Andreasen, M.M., C.T. Hansen, and N.H. Mortensen. *The Structuring of Products and Product Programmes*. In *2nd WDK Worshop on Product Structuring*. 1996. Delft University of Technology, Delft, The Netherlands.

[4]   Erens, F., Verhulst, K. *Architectures for product families*. In Proceedings of the 2nd WDK Workshop on Product Sructuring. June 3-4, 1996. Delft University of Technology. The Netherlands. pp. 45-60

[5]   Riitahuhta, A.; Andreasen, M.M.; "Configuration by Modularization". In: Proceedings of NordDesign'98. Stockholm 1998.

[6]   Borowski K-H. Das Baukastensystem in der Technik. Springer -Verlag, Berlin, Göttingen,Heidelberg. 1961.

[7]   Pimmler T.U., Eppinger S.D., Integration analysis of product decompositions. In: *DE-Vol. 68, Design Theory and Methodology – DTM '94*, ASME. 1994

[8]   Reinders H., A Tool for Diversity Management during Product-Development. CFT-technology. Report 24/89EN, Philips Centre for Manufacturing Technology, 1998

[9]   Mørup, M. *Design for Quality*, Diss. Institute for Engineering Design, Technical University of Denmark, 1993.

[10]Steward D., The Design Structure System: A Method for Managing the Design of Complex Systems, IEE Transactions on Engineering Management, Vol.Em- 28, No. 3, August 1981. 71-74. , 1981.

A Framework for Evaluating Commonality

Roger B. Stake

Royal Institute of Technology
Department of Manufacturing Systems
Div. of Assembly Systems
S-100 44 Stockholm, Sweden
e-mail: rst@cadcam.kth.se

Abstract. The issue of product families and product platforms have gained an increased interest both in academia and industry. The concept of commonality is often mentioned within this framework. However, commonality may occur at different levels of abstraction, commonality of functions or parts, and at different levels of detail, commonality of modules or parts. This paper presents a literature survey of related research on product families, product structuring and measurements associated to those areas, in order to clarify what to consider when measuring commonality. The result is a framework for measuring commonality. Firstly, there is a need to clarify the kind of commonality measured, that is, on what architectural and structural level (level of abstraction and level of detail) is the commonality measured. Secondly, there is also a need to clarify what reference that is used for calculating the commonality. The concepts of product family commonality and product variant commonality are introduced to distinguish the reference for calculating the commonality.

1 Introduction

Product variety and commonality are often discussed within the framework of modular products and product families. Modularity is often discussed in terms of having the ability to create product variants by altering some modules in the final product variant or by having a modular system that allows a combinatorial assembly of the final product variant.

An increasing number of companies see component sharing (commonality) as a way to have high variety in the marketplace and low variety in the plant. In the 1980s, Black & Decker rationalised its product lines by clustering its products by motor sizes. By eliminating unnecessary proliferation, the number of motor sizes was reduced five fold, despite an increase in the number of end products [1]. Products which share more parts and modules within a product family achieve greater inventory reductions, exhibit less part variability, improve standardisation, and shorten development lead times because more parts are reused and fewer new parts have to be designed [25].

However, commonality may be considered in a number of different ways, ranging from pure standardisation of parts to the re-use of conceptual solutions within a product or a product family. Commonality is also one of the factors that may be used for evaluating product families. Hence, there is a need to identify what to consider when measuring commonality.

The objective of this paper is to explore the issue of commonality in order to clarify what different kinds of commonality there are, that is, provide a framework for evaluating commonality. The rest of the paper is outlined as follows: section 2 gives an introduction to product families, section 3 explores the issue of commonality, section 4 presents several measurements associated with product families and commonality and section 5 concludes the paper by presenting a framework for evaluating commonality.

2   Product Family Fundamentals

2.1
Product Platforms and Product Families

Product families and product platforms have been discussed frequently in both academic and industrial literature. Numerous examples of successful product families have been given in different market segments. Some examples of product families are:

- Automotive industry – Scania trucks [2], Nippondenso [3], VW A4[4]
- Consumer goods – Swatch watches [5], Sony [6].
- Office equipment industry – Xerox [5], Hewlett Packard Printers [7]
- Power tools – Black & Decker – Universal Motor Platform [8]
- Airspace industry – Rolls-Royce – Aircraft Engine Platforms and Boeing 747-Series [9]

Product families are often planned so that a number of derivative products can be efficiently created from the foundation of a common core technology, called the product platform. The platform is a set of subsystems and interfaces that form a common structure from which a stream of derivative products can be efficiently developed and produced [1]. A product family may moreover be defined as a set of products, described by a parameterised data-structure, where a unique variant of the family is defined if all parameters have a value [10]. Furthermore, it may be described as products with identical internal interfaces, that is, interfaces between the products components for all variants in each of the functional, technology and physical domain. The interfaces must be standardised in each of the domains to allow the ability of full exchange of components [11]. In a similar context, the term modular system could be used for the total set of modules creating the product variety [12, 13].

These statements of product platforms and product families are focused on the actual product. A broader view of the product platform concept in product development is that the product platform may be seen as a collection of assets which are shared by a set of products. These assets can be divided into four categories [14]:

- Components – the part design of a product, the fixtures and tools needed to make them, the circuit designs, and the programs burned into programmable chips or stored on disks.
- Processes – the equipment used to make components or to assemble components into products, and the design of the associated production process and supply chain.

- Knowledge – design knowledge, technology applications and limitations, production techniques, mathematical models and testing methods.
- People and relationships – teams, relationships among team members, relationships between team and larger organisations and relations with a network of suppliers.

In order to elaborate more on the issue of product platforms further definitions are needed. In [15] the following, somewhat more abstract or general, definitions of product platforms are given

- A product platform ... encompassing the design and components shared by a set of products [16]
- A platform is the physical implementation of a technical design that serves as the base architecture for a series of derivative products [17]
- The platform ... is a collection of common elements, especially the underlying core technology, implemented across a range of products [18]
- A product platform is the foundation for a number of related products, typically a product line. While all products are unique in some way, they are related by the common characteristics of the product platform [18]
- A platform is a relatively large set of products' components that are physically connected as a stable sub-assembly and is common to different final models [19]
- The set of assets shared by different vehicles. These assets include at least the familiar parts of the car (such as the main stampings in the engine compartment) and the machinery, equipment and tooling in the assembly plants [20]
- The process of identifying and exploiting commonalties among the firm's offerings, target markets and the process for creating and delivering offerings [21]

The words "common", "shared" and "base" occur often in all of the definitions above. Hence, the platform may be regarded as something that is common for a range of products; the common elements may be parts or components of the product. However, the product structure, the processes in which the products are produced, or the interfaces may also be regarded as the product platform while they are common for a range of products.

In conclusion, there are two different viewpoints of product platforms:

- *The commonality based viewpoint* – regards the common elements of the product family to be the product platform
- *The resource based viewpoint* – regards all the resources necessary to build the product family to be the product platform

2.2
Benefits of Product Families and Modular Products

There exist many benefits of product families and modular products. Several of these benefits are derived from the fact that some parts are used in several product variants. The benefits can be divided into categories largely according to the product life cycle: product development and design, variance, production, quality, purchasing and after-sales. The topic of modularity is often discussed in the context of variety. One common remark is the explosion of variants and the problem that it inflicts to the production. Modularity reduces these problems by increasing the degree of commonality, while allowing the consumer to mix and match elements to come up with a final product that suits their preferences and needs [22].

The function of a component may be generic enough so that a standard interface protocol can be adopted and identical components can be used in more than one type of product variants. The definition of functionality and specification of uncoupled interfaces enables a black box approach [23]. The effect of out-sourcing is often a concentration in the core competence of the company; they do what they are best at. The ability to decide what to manufacture and what to buy could be considered to be the real core competence [24].

Standardised modules are likely to be allocated in many product variants. This allows for economy of scale by using the same module in many product variants (commonality), that is, a common module may be produced in large volumes [11]. Collier [25] concludes that component part commonality can have a significant impact on system performance, reducing manufacturing costs, total costs, and delivery performance.

Modules may also intentionally be reused in many generations, where only some parts are updated [26]. This also allows for concentration of technical risk to a few modules that increases the probability of success for the whole system [27].

Commonality can also be used in the product development to reduce time and cost by re-use the design of modules (concepts or detail drawings) between different product variants, either in the present range or in future derivatives. Rationalisation is provided if the particular function variant is based on a combination of fixed individual parts and or assemblies [12]. The availability of modules for re-use also makes the product development projects more predictable [28]. Commonality will likely lead to economies of scale in component production, enhanced performance through continuos refinement, broad amortisation of development costs and reduced materials management costs [23].

Since commonality or standardisation is the foundation for many of the benefits of product families, they are important issues to consider when developing a product family.

3   Commonality

3.1
What is Commonality

According to Collins Cobuild Dictionary commonality is: *"two or more things have something in common if they have the same characteristic or feature"* and according to Webster dictionary: *"commonality is possession of common features or attributes*". So, commonality is related to the fact that two or more objects share some characteristics.

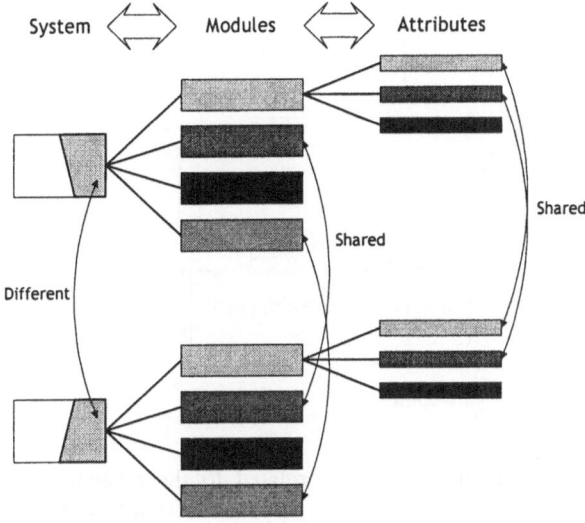

Fig. 1. Sharing at different levels [28]

However, as Fig. 1 illustrates, the sharing may be that different systems share the same modules or different modules share the same attributes. The question is to decide what characteristics the objects should share, that is, commonality could occur at different levels of abstraction and detail of the product. Hence, the issues of product structuring and product architecture are essential for understanding the concept of commonality.

3.2
Product Structuring – the Essence of Product Platforms

Today, a one-off product is rarely developed; more often a range of products is developed based on the same type of technology. In order to use this fact efficiently, a modular system may be created to reap some of the benefits mentioned earlier. Arranging the element and interfaces in an appropriate product structure does this. For example, the structure of a product could be described as the way in which its elements are interrelated in a system model, based on the actual viewpoint which may be functional, parts oriented, kinematics, etc. [29]. The structure could also be expressed as the set of elements in a system and the set of relationships that connect these elements to one another [30]. The structuring may be done during the development at various levels of abstraction, for example, function, working, construction and system structure [12] or process, function,

organ and component structure [30]. Fig. 2 illustrates a product structure based on
the viewpoint of parts.

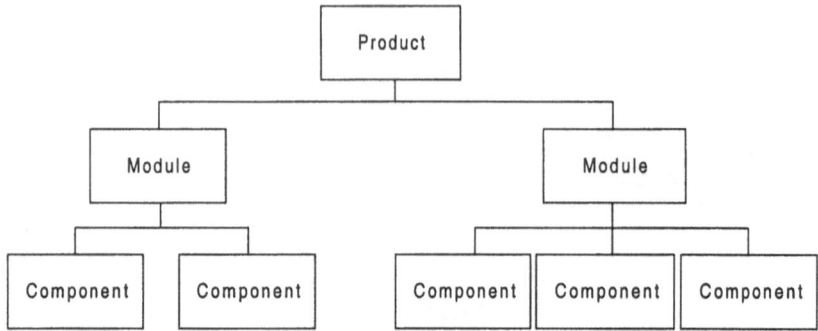

Fig. 2. A product structure from a parts viewpoint

The nature of the product structure is closely related to the design of the
product. The shape of the product part structure is partly determined by the
number of components and parts used at each level, the more used the wider the
shape. Standardisation of components in order to reduce the number of parts will
slim the shape of the product part structure [31]. In Fig. 3, four different shapes of
product part structures are represented.

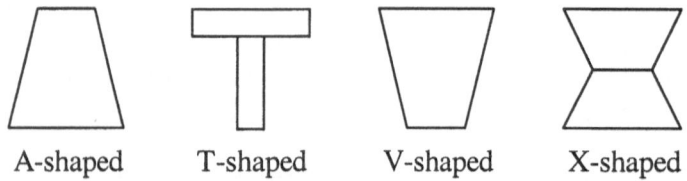

A-shaped   T-shaped   V-shaped   X-shaped

Fig. 3. Different shapes of product structures [31]

Other authors deal with the same issue by using the expression product
architecture, expressed as the scheme by which the function elements of the
product are mapped onto physical components or chunks [23, 5], respectively.

Colloquially, structure and architecture are used synonymously, with the
distinction that the definition for product structure seems not to include the
relations between the levels of abstractions as in the definitions for architecture. In
conclusion, the structural level described the level of detail and the architectural
level the level of abstraction.

Nevertheless, when a product is expanded to a product family, the physical
structure should be superimposed by principles for variation and familiarities seen

from the product life cycle activities [32]. Consequently, products should be designed based on a structure that could make use of the familiarity between products, normally referred to as a modular structure. Modularity is often referred to in the context of product structuring, where modularity means a structure that exhibits certain characteristics. These characteristics may be describes as the type of modular systems that is used.

3.3
Types of Modular Systems

Typically, modular products may be grouped into three types of modular systems, which are slot, bus and sectional modularity [23], illustrated in Fig.4. In the sectional system, the product variants are configured rather freely from modules that can be combined in several manners by standardised interfaces. In bus modularity, the modules have standard interface that allows attachment of various modules in various positions to a base module. Finally, slot modularity refers to a system where each type of module is connected in a certain position by a standard interface.

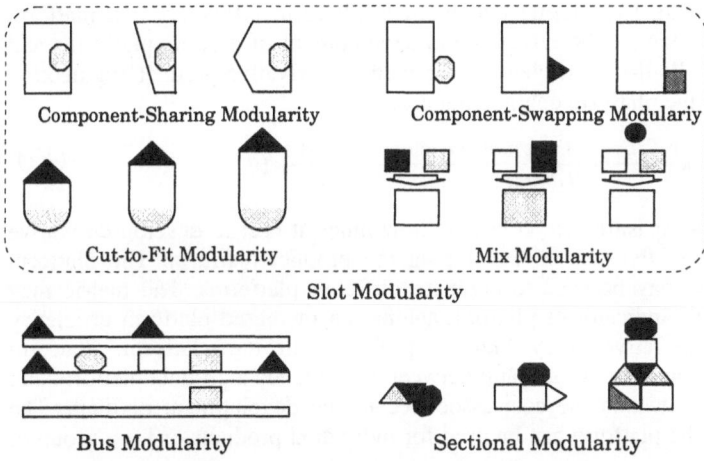

Fig. 4. Types of modularity, adopted from [32, 23]

The different types of modularity may also be expanded to more categories also based on how product variety is created. Product variety can be created in different stages of the production process, that is, customisation in design, fabrication, assembly or even after production [33]. In [33] it is also suggested that re-use may be applied at different levels, that is re-use of elements ranging from principles down to parts, even elements such as manuals are considered to be sources for re-use.

3.4
Commonality Versus Re-use

Re-use is often mentioned in the context of product families and modular products. However, re-use shows similarities with commonality. The common aspect is that something is used in several entities and the differentiating aspect is time. Commonality is the use of something unchanged in several entities in the present and re-use is the use of something unchanged in the future.

4   Product Family and Commonality Measurements

4.1
Product Family Measurements

Various researchers have presented metrics for measuring either the efficiency or effectiveness of a product platform or the degree of commonality within a product family. Meyer et al. [34] define two metrics, platform efficiency and platform effectiveness, to manage the research and development of product platforms and product families. Platform efficiency is defined as derivative product engineering costs divided by the platform engineering cost.

$$Platform\ efficency = \frac{Derivative\ product\ engineering\ costs}{Platform\ engineering\ costs} \cdot 100 \qquad (4:1)$$

The efficiency measurement considers how much it cost to develop derivative products related to the cost of developing the product platform. The platform efficiency metric may be used to compare different platforms. The metric may also be used as an indicator of platform ageing; an increased platform efficiency of successive derivatives may indicate problems in the platform. Platform effectiveness is calculated as the ratio between the revenue a product platform and its derivatives create and the cost associated to the development of them. The effectiveness of the platform can be used for individual products or for a group of products.

$$Platform\ effectivness = \frac{Net\ sales\ of\ a\ derivative\ product}{Development\ costs\ of\ a\ derivative\ product} \cdot 100 \qquad (4:2)$$

These metrics require extensive information regarding costs and revenues for a product family. This information is often only known when the product platform and its derivatives have been developed, produced and offered to the market.

Furthermore, these measurements are difficult for designers to use during the conceptual and the design phases of the development process, all the information needed for the calculation is not present. Some more pragmatic metrics, that do not need the same amount of information, have also been proposed for the evaluation of product families.

Sundgren [15] presents a product platform utilisation measurement where alternative product platforms may be evaluated by comparing the number of theoretical possible product variants with the number of actual offered product variants.

$$Platform\ utalisation = \frac{\#\ of\ utalised\ product\ var\,iants}{theoretical\ \#\ of\ possible\ product\ var\,iants} \cdot 100 \qquad (4:3)$$

Similarly, flexibility may be measured as the offered product variants in comparison to the wanted product variants in the market [13]. Erixon [35] presents a measure of the variant flexibility as measured by dividing the number of product variants with the total number of modules, giving a value of the yield of the modular system; a higher value indicates greater leverage of the system.

$$\text{Variant flexibility}\ E_{var} = \frac{N_{var}}{N_{mtot}} \qquad (4:4)$$

where:
N_{var} = Total number of product variants
N_{mtot} = Total number of modules in product family

Erixon [35] also presents a measurement of the assortment complexity (product family complexity) that is derived from Pugh's complexity measurement [36].

$$\text{assortment complexity} = \sqrt[3]{N_m \cdot N_{mtot} \cdot N_c} \qquad (4:5)$$

where:
N_m = Number of modules in one product (average)
N_{mtot} = Total number of modules in product family
N_c = Number of contact surfaces in interface

4.2
Commonality Measurements

Instead of the above-presented metrics, commonality could also be used to evaluate product families. The benefit of using commonality is that it may be measured at different levels of detail and abstraction. In the early design phases, commonality among principles may be measured and in the later design phases the degree of commonality among parts could be calculated. Several commonality indices have been proposed for measuring the degree of commonality within a product family.

Maier [37] presents an index for degree of similarity, calculated by dividing the number of similar parts with the total number of parts. The index falls between 0 and 1, where 0 characterises divergent designs, 0,25 similar design and 1 is identical design.

$$CI = \frac{\sum \text{number of similar parts}}{\sum \text{total number of parts}} \qquad (4:6)$$

Martin et al. [38] presents a similar index, called the commonality index. The index is normalised to range from 0 to 1 and is calculated as the number of unique parts divided by the total number of parts. A smaller number indicates a better index, an index of 1 indicates the worst possible case. The commonality index presented by Martin et al. [38] is, compared with the index presented by Maier [37], more of a uniqueness index. That is, the index assesses the degree of unique parts.

$$CI = \frac{u}{\sum_{j=1}^{v_n} p_j} \tag{4:7}$$

where:
u = number of unique part numbers
p_j = number of parts in model j
v_n = final number of varieties offered

Later, Martin and Ishii [39] have reformulated the commonality index to be more of a commonality index instead of the former uniqueness index.

$$CI = 1 - \frac{u - \max p_j}{\sum_{j=1}^{v_n} p_j - \max p_j} \tag{4:8}$$

where:
u = number of unique part numbers
p_j = number of parts in model j
v_n = final number of varieties offered

A higher CI is better since it indicates that the different product variants within a product family are being produced with fewer unique parts. A comment is that the index is not applicable for product families consisting of one product variant, if one could call such case a product family. In such case, the maximum number of parts in a model equals the sum of all parts in all product variants and consequently the denominator will equal zero.

Collier [25, 40] presents the Degree of Commonality Index, which is a measure based on the bill of materials (BOM) of the product. The index can be applied to different levels of the product, for example, a single product variant or the whole product family. The lower bound on the degree of commonality is one (no commonality). The upper bound on the degree of commonality is β (complete commonality). Complete commonality results when the total number of distinct components (d) equals one. The index is the inverse of the index presented by Martin and Ishii [38]

$$\text{Degree of commonality index } (C) = \frac{\sum_{j=i+1}^{i+d} \Phi_j}{d} \quad 1 \le C \le \beta \quad \beta = \sum_{j=i+1}^{i+d} \Phi_j \tag{4:9}$$

where:

Φ = the number of immediate parents component j has over a set of end items or product structure level(s)

d = the total number of distinct components in the set of end items or product structure level(s)

i = the total number of end items or the total number of highest level parent items for the product structure level(s)

β = the total number of immediate parents for all distinct component parts over a set of end items or product structure levels

The presented indices are based on the number of parts of a product or product family. Of course also parts, modules or subsystems could be used in the equations to assess the degree of commonality among them. The following commonality indices assess more factors than just parts when calculating the degree of commonality.

Kota et al. [41] present a product line commonality index (PCI):

$$PCI = \frac{\sum\limits_{i=1}^{P} CCI_i - \sum\limits_{i=1}^{P} Min\ CCI_i}{\sum\limits_{i=1}^{P} Max\ CCI_i - \sum\limits_{i=1}^{P} Min\ CCI_i} \qquad PCI = \frac{\sum\limits_{i=1}^{P} n_i \cdot f_{1i} \cdot f_{2i} \cdot f_{3i} - \sum\limits_{i=1}^{P} \frac{1}{n_i^2}}{\sum\limits_{i=1}^{P} n_i - \sum\limits_{i=1}^{P} \frac{1}{n_i^2}} \cdot 100 \qquad (4{:}10)$$

where:

CCI_i = Component commonality index for component i

$Max\ CCI_i$ = Maximum possible component commonality index for component

$Min\ CCI_i$ = Minimum possible component commonality index for component

P = Total number of non-differentiating components that can potentially be standardised across models

n_i = number of products in the products family that have component

f_{1i} = Part size and shape factor for component

f_{2i} = Material and manufacturing processes factor for component

f_{3i} = Parts assembly and fastening schemes factor for component

$$CCI_i = n_i \cdot f_{1i} \cdot f_{2i} \cdot f_{3i} \qquad Max\ CCI_i = n_i \qquad Min\ CCI_i = n_i \cdot \frac{1}{n_i} \cdot \frac{1}{n_i} \cdot \frac{1}{n_i} = \frac{1}{n_i^2}$$

The PCI is expressed as a percentage and assumes the values between 0 and 100. A PCI score of 0 indicates that none of the non-differentiating parts are shared across models and a score of 100 indicates that all of the non-differentiating parts are shared across models and they are all identical in size, shape and made by using the same materials and manufacturing processes and the fastening methods used are identical across all models.

Furthermore, [42] measures the platform architecture commonality by calculating the component commonality, connections commonality, and assembly component loading commonality and assembly workstation commonality. All these measurements are calculated by dividing the number of common parts with the total number of parts. The platform architecture commonality is then calculated by sum all the different commonality measurements.

$$Commonality = \sum_{i=1}^{3} I_i \cdot C_i = I_c \cdot C_c + I_n \cdot C_n + I_a \cdot C_a \qquad (4:11)$$

where

I_i = importance (weighting factor), normalised, that is, $\Sigma I_i = 1$

C_i = % commonality measures

A comment is that Siddique [42] presents four different components but only involves three of them when calculating the aggregated commonality of the product platform.

4.3
Reflections on the Different Measurements

The measurement for assessing the degree of commonality may be divided into two different categories, pragmatic measures and analytical measures.

Table 1. Classification of the different measurements

Pragmatic	Analytical
Maier [37]	Kota [41]
Siddique [42]	Collier [25, 40]
	Martin and Ishii [38]
	Martin and Ishii [39]

The pragmatic measures are those defined as the amount of common parts divided by the total number of parts. The reason for calling them pragmatic is that the definition of part is not clear, it could be a subassembly, a component, a module or even as vague as an attribute. These measures may also be used in the earlier phases of the product development phase where the whole product is not defined.

The analytical measures are those that take the whole product structure into account. These measures need lots of information, preferably should the product be completely designed and ready for manufacturing.

5   Conclusion and Discussion

The amount of commonality can be measured in several different ways, as presented in chapter 4.2. Considering the earlier presented discussion there are a need to clarify what kind of commonality that is measured.

First of all commonality can be measured at different levels of abstraction, that is, either could functions be used in several product variants or could parts be shared. Secondly, commonality could also be measured as the amount of modules used in several product variants or as the number of parts shared. Therefore, there is a need to clarify what architectural and structural level (level of abstraction and level of detail) the commonality is measured at.

The kind of commonality can be divided into the following two categories:
- *Product architecture level* – the degree of commonality may be considered for functions or parts
- *Product structure level* – the degree of commonality may be considered for modules, components or parts

There is also a need to clarify the reference used for calculating the degree of commonality. Either could one product variant be used as reference or the whole product family.
- *PFC – Product family commonality* – the degree of commonality within the product family
- *PVC – Product variant commonality* – the degree of commonality within a single product variant.

The actual calculation of the commonality may be performed in many different ways; chapter 4.2 provides a sample of measurements that could be used. In the following exemplification of the framework the commonality index presented by [37] will be used, that is, the number of similar parts divided by all parts. According to the framework, the kind of commonality should be specified. In this example the commonality for the architectural level of parts and for the structural levels of modules and components will be calculated. The fictive product structure is depicted in

Fig. 5. The product is built of two modules that consist of three components each. Furthermore, some of the components exist in variants.

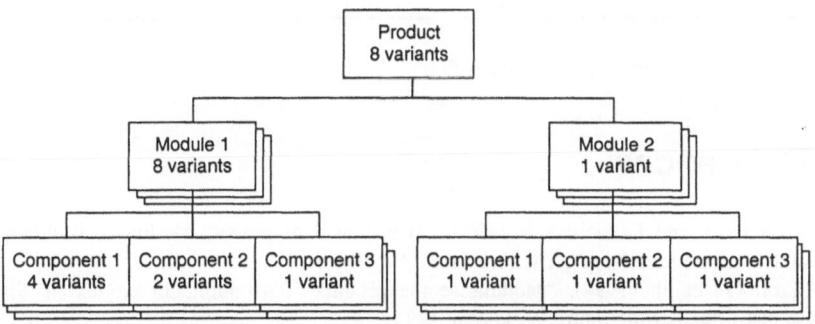

Fig. 5. The fictive product structure used for exemplifying the framework

The PFC at module level is calculated as, 1 out of a total of 9 modules (11%) are common within the product family. The PVC is calculated as, 1 out of 2 modules (50%) are common for several product variants within the product variant. The commonalties at the structural level of components may also be calculated in the same manner. Resulting in, PFC=4/10 (40%) and PVC=4/6 (67%). The different commonalties are summarised in Fig. 6.

Figure 6 illustrates the importance of specifying the kind of commonality and the structural level used for measuring commonality, as well as the reference used for the calculation. Only this simple example results in four different commonality measurements.

If commonality is to be used for comparing or evaluating product families, the framework could be used to ensure that the same commonality is measured and compared. It also provides a mechanism for expressing different kind of commonalties throughout the development process. In early design phases a suitable measure may be the degree of commonality of principle solutions and in the later phases may the degree of commonality of parts be more appropriate, that is, different architectural levels. Commonality at different structural levels may be used to identify how different stages of the production process may benefit. As an example, sales and product development may benefit from high degree of commonality of modules, while manufacturing may benefit from a high degree of commonality of parts.

		Degree of commonality for architectural level of parts	
		PVC	PFC
Structlevural l	Module	1/2 = 50%	1/9 = 11%
	Component	4/6 = 67%	4/10 = 40%

Fig. 6. The different commonalties exemplified

6   References

[1]   Meyer M.H. and Lehnerd A.P. (1997), "The power of product platforms, Building value and cost leadership", The Free Press, New York, 1997
[2]   Erixon G., et al. (1994), "Modularise the product" ("Modulindela produkten" in Swedish), Industrilitteratur, Stockholm, 1994
[3]   Whitney D.E. (1993), "Nippondenso Co. Ltd.: A case study of strategic product design", Research in Engineering Design, Vol. 5, 1993
[4]   Bremner R. (1999), "Cutting edge platforms" Financial Times, Automotive World, September, 1999
[5]   Ulrich, K.T. and Eppinger S.D. (1995), "Product design and development", McGraw-Hill, New York, 1995
[6]   Sanderson, S.W. & Uzumeri, M. (1997), "The Innovation Imperative: Strategies for Managing Product Models and Families", Irwin, Chicago, 1997
[7]   Lee H.L. & Tang C.S. (1993), "Modelling the cost and benefits of delayed product differentiation", Management Science, Vol. 43, No. 1, 1993

[8] Lehnerd A.P. (1987), "Revitalizing the manufacture and design of mature global products", in Guille B.R. and Brooks H., eds., "Technology and global industries", National Academy Press, Washington, 1987

[9] Rothwell, R. & Gardiner, P. (1990), "Robustness and Product Design Families", Design Management: A Handbook of Issues and Methods (Oakley, M., ed.), Basil Blackwell Inc., Cambridge, 1990

[10] Wortman H.C. & Erens F.J. (1995), "Control of variety by generic product modelling", 1:st World Congress. on Intelligent Manufacturing Processes and Systems., San Juan, Puerto Rico, 1995

[11] Erens F. & Verhulst K. (1996), "Architectures for product families", 2nd WDK-workshop on Product Structuring, TU Delft, The Netherlands, June 3-4, 1996

[12] Pahl G. & Beitz W (1996), "Engineering Design – a systematic approach", Springer-Verlag, London, 1996

[13] Kohlhase N. & Birkhofer H. (1996), "Development of modular structures: the prerequisite for successful modular products", Journal of Engineering Design, Vol. 7, No. 3, Sept., p. 279-291, 1996

[14] Robertson D. and Ulrich K. T. (1998), "Planning for product platforms", Sloan Management Review, Summer, p. 19-31, 1998

[15] Sundgren N. (1998), "Product platform development – Managerial issues in manufacturing firms", Doctoral thesis, Chalmers university of Technology, Dept. of Operations management and work organisation, 1998

[16] Meyer M. & Utterbeck J. (1993), "The product family and the dynamics of core capability", Sloan Management Review, Spring, p. 29-47, 1993

[17] Meyer M. & Lopez L. (1995), "Technology strategy in software products company", Journal of Product Innovation Management, 12, p. 294-306, 1995

[18] McGrath M. (1995), "Product strategy for high-technology companies", Irwin Professional Publishing, New York

[19] Muffatto M. (1997), "Enhancing the product development process through a platform approach", Proceedings of the 10th International Conference in Production Research, Osaka, 4-8 August, 1997

[20] Ericsson J, et al. (1996), "Sharing parts across car models: lessons from the manufactures", Europe's Automotive Components Business, 1st quarter, p. 150-171, 1996

[21] Sawhney M.S. (1998), "Leveraged high-variety strategies: from portfolio thinking to platform thinking", Journal of the Academy of Marketing Science, Vol. 26, No. 1, p. 54-61, 1998

[22] Baldwin C.Y. & Clark K.B. (1997), "Managing in an age of modularity", Harvard Business Review, September - October, 1997

[23] Ulrich, K.T. (1995), "The role of product architecture in the manufacturing firm", Research Policy, Vol. 24, p. 419-440, 1995

[24] Fine C.H. & Whitney D.E. (1996), "Is the make-buy decision process a core competence", MIT Center for technology, policy and industrial development, Feb. 1996

[25] Collier D.A. (1981), "The measurement and operating benefits of component part commonality", Decision Sciences. Vol. 12, No. 1, p.85-96, 1981

[26] Sanchez R. (1994), "Towards a science of strategic product design", 2nd International product development management conference on new approaches to development and engineering, Gothenburg, Sweden, p.564-578, 1994

[27] Smith P.G. & Reinertsen D.G. (1995), "Developing products in half the time", Van Nostrand Reinhold, New York, 1995

[28] Fujita K., et al. (1998), "Simultaneous optimisation of product family sharing system structure and configuration", ASME Design Engineering Technical Conferences, September 13-16 1998, Atlanta, USA., DETC98/DFM-5722

[28] Reinertsen D.G. (1992), "Use product architecture to slash design time", Electronic Design. Vol.40, No.25, 3 Dec., p.59-62, 1992

[29] Andreasen M.M. (1995), "On structure and structuring", Workshop Fertigungs-gerechtes Konstruieren, Erlangen, Oct. 1995

[30] Hubka V. & Eder W.E. (1988), "Theory of Technical Systems", Springer-Verlag, Berlin, 1988

[31] Slack N., et al. (1998), "Operations management" Second edition, Pitman Publishing, London, 1998

[32] Andreasen M.M., et al. (1996), The structuring of products and product programmes, Proc. 2nd WDK workshop on product structuring, Delft, June 3-4, 1996

[32] Ulrich, K.T. and Tung K. (1991), "Fundamentals of product modularity", ASME winter meeting symposium on issues in design/manufacturing integration, Atlanta USA, 1991

[33] Miller T.D. & Elgaard P. (1999), "Structuring principles for the designer". the International CIRP Design Seminar, 24-26 Mar. 1999, University of Twente, The Netherlands",

[34] Meyer M.H., Terzakian P. & Utterback J.M. (1997), "Metrics for managing research and development in the context of the product family", Management Science, Vol. 43 No. 1, 1997

[35] Erixon G. (1998), "Modular Function Deployment – A method for product modularization", Doctoral thesis, Royal Institute of Technology (KTH), Dept. of Manufacturing system, Sweden, 1998

[36] Pugh, S. (1991), "Total Design – Integrated Methods for Successful Product Engineering", Addison-Wesley Publishing Company, New York, 1991

[37] Maier T. (1993), "Similarity information through commonality analysis of a product program", (in German), ICED, The Hague, The Netherlands, 1993

[38] Martin M.V. & Ishii K. (1996), "Design for variety: a methodology for understanding the costs of product profileration", ASME, 18-22 Aug., Irvine, USA, 1996

[39] Martin M.V. & Ishii K. (1997), "Design for variety: development of complexity indices and design charts", ASME, 14-17 Sept., Sacramento, USA, 1997

[40] Collier, D.A. (1982), "Aggregate Safety Stock Levels and Component Part Commonality", Management Science, Vol. 28, No. 22, p. 1296-1303, 1982

[41] Kota S. & Sethuraman K. (1998), "Managing variety in product families through design for commonality", ASME, 13-16 Sept., Atlanta, USA, 1998

[42] Siddique Z., Rosen D.W. & Wang N. (1998), "On the applicability of product variety design concepts to automotive platform commonality", ASME, 13-16 Sept., Atlanta, USA, 1998

Part IV

Supporting Modeling and IT-Tools

Conclusions and Discussion for Further Research

Antti Pulkkinen

Tampere University of Technology
P.O.Box 589
33101 Tampere, Finland
e-mail: pulkkine@ruuvi.me.tut.fi

Abstract. Remarks on the state of Product Structuring research are given. A framework on directions and levels on Modular Engineering is presented. Demands and wishes for future research topics are listed.

1 Introduction

This paper is based on the records made in the concluding section of the workshop. Marcel Tichem led the section of the workshop from Delft University of Technology. It started by noting the results from the workshop and continued by suggesting topics of necessary research topics in the area. Furthermore, each participant was suggested to give two opinions about the interesting research topics. In the following pages, an interpretation of the discussion is presented.

2 Discussion about the Workshop Results

Academics and researchers (9 presentations out of 12) mainly gave the presentations. A consultant gave one presentation and only two presenters came from purely industrial background. Thus the workshop was more an academic review on state-of-the-art than state-of-the-practise. To get a perception of the industry needs there was an introduction by industrialists in two sections (development of product portfolios, modelling and IT-support). Despite this, it was pointed out that more industry input should arise in the future workshops.

Nowadays there exist many methods and criteria. These methods are suggested to help for Modular Engineering (ME) and Design for Configuration. Some of them were examined in the third section of the workshop. Furthermore, tools and IT-support, which were the topic of fourth section of the workshop, are developed for easing the application and implementation of ME.

In ideal situation, research on theory should support research and development of methods and tools. This does not seem to be the case in product structuring research, but different sections seem to be rather independent of each other. An indication of this is the fact that in the workshop the contents of different sections do not seem to meet. The research on methods and criteria do not seem to support the development of tools and IT-support. Applying these methods and tools in the analysis of customers, markets and technology does not seem to happen. Moreover, development of re-usable (knowledge, activity and product) structures does not seem to have methodical support.

Also the position and dimensions of ME (and Product Structuring) is unclear in two ways. First, it is unclear how to adopt ME in practise and second it is unclear what elements of research are needed. From industrial point of view this leads to a situation where a decision of applying a certain method is not well founded. From academic point of view, the positioning of a research effort is difficult. In the Fig. 1. there is a one suggestion of the needed directions and levels in Product Structuring research.

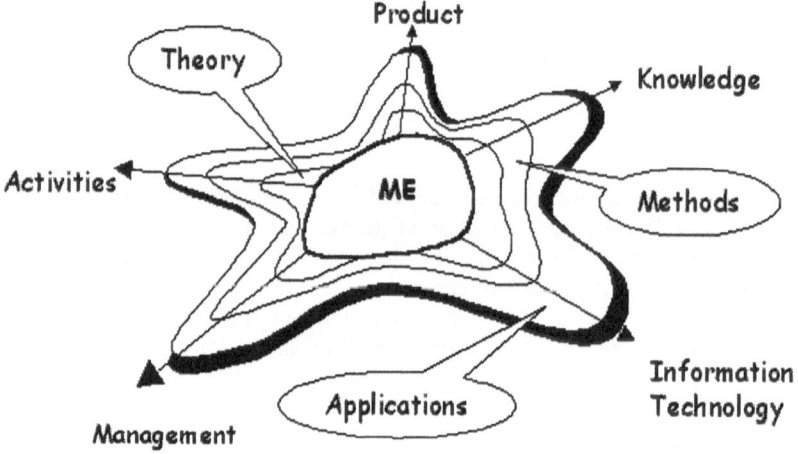

Fig. 1. Directions and levels of ME research

Modular Engineering is in the center of the Fig. 1. The effects of ME are manifested in activities, product and knowledge structures. In practice, both management and information technology have an effect on the application of ME. Each of the directions have different levels on research depending on the generality and applicability. In the figure these levels are represented with "growth rings". E.g. research on product / artefact theory should support the development of new methods. In order to obtain general, valid and implementable methods and tools, *research on connecting the different levels of research is needed.*

Generally every research project has to be focused to secure the results. For instance, many concurrent engineering and engineering design research projects concentrate on different product life cycle phases.

However, it is hard to find a product development project that has its goal on improvements in a single life cycle phase. Also in ME many product life cycle phases have to be taken into account. For instance, in one case it was said, "a module should fit into a container". This example points out a requirement from logistics for product structuring.

Configuration may occur either in sales (by using front office / sales configurator) or in engineering (by using back office / engineering configurator).

In any case, the objective is to determine an individual configuration and produce equivalent product individual for a specific customer. Thus, the decisions in Design for Configuration have an influence on the performance of engineering and sales in a company. Additionally, production rationalization is usually being pursued. Information technology is a way for putting sustainable methods and applications to use (i.e. to attain the performance effects and rationalization).

Product life cycle phases are not isolated issues in practical Product Development. Similarly, Product Structuring directions are not isolated issues in reality. Some of the research is concentrated on finding the metrics (like commonality indices) that represents the Product Structuring virtues. However, in this kind of research careful attention should be paid on the relation between different directions of product structuring (i.e. what is stake and result). In order to obtain general, valid and implementable methods and tools, *research on connecting the different research directions is needed.*

3 Further Research

The participants were asked to give their future interest in certain research topics. The following table is a transcript of this.

Table 1. Future research topics

	Topics	Interest among participants
1.	Introduction of ME in companies	♪♪♪♪♪ ♪♪♪
2.	Handling of product & (design) process related data of product families	♪♪♪♪♪ ♪♪♪
3.	Design methods that make ME an integral part of the PDP (Product Development Process)	♪♪♪♪♪ ♪
4.	Decision making in ME	♪♪♪♪♪ ♪
5.	ME in relation to re-use	♪♪♪♪♪
6.	Operational orgnisation of ME	♪♪♪
7.	In company case studies on ME	♪♪♪
8.	Evaluation methods and metrics for ME	♪♪♪
9.	Terminology	♪♪
10.	(IT) tools for ME support	♪♪
11.	Formal modelling of Product Structures / Families	♪♪

Table 1 presents the participants impression on the most interesting / necessary research topic on the area. The order of topics represents how many participants considered the topic to be of interest in the year 2000. E.g. Evaluation methods and metrics for ME interested two participants. Every participant was asked to give two different topics of his/her interest.

Putting ME into use is interesting for most of the participants. Since applying ME in industry by means of IT (i.e. topics 1. and 2.) seems to lead the future research topic "wish list". Topics 3. and 4. are related to product development issues. These and some others (6. and 8.) are in the level of research on methods. However, research on theory base (topics 5., 9. and 11.) and supporting tools (10.) seems to interest the least number of participants.

Table 1. does not express the relation of the participants' choices (two choices were made). Thus, no conclusion can be drawn on how the future research will bind the levels and directions of Product Structuring and Modular Engineering.

4 Conclusion

Based on the papers and discussions several questions were found. The industrial interest on Product Structuring issues exists. However, industrial participation should be enhanced. An effort to this direction was taken in the workshop. The position of Modular Engineering and Product Structuring among Engineering Design is unclear. This holds true both in research and in practice.

A framework on relating the directions and levels of research topics is given. Based on the framework, it is suggested that two kinds of research should be carried out:
- research on connecting the different levels (theory, methods, applications) of Product Structuring research
- research on connecting the different Product Structuring research directions (products, knowledge, activities, information technology, and management)

The list of future research topics is given. It suggests that there is a practical need on Product Structuring research and development. Yet, rather limited enthusiasm on building a solid theory base exists.

Preservation of Engineering Knowledge in Configuration Systems

Thomas Jensen

Danish Technological Institute
Center for Production
Gregersensvej, P.O.Box 141
DK-2630 Taastrup
e-mail: thomas.jensen@teknologisk.dk

Abstract. It is widely recognized that knowledge is an important asset for most companies. During the last decade, knowledge structuring has gained interest by industry with configuration systems as one of the most widespread means. Commercial configuration systems support modular configuration, where a product can be configured by composing a set of predefined modules. The core of a configuration system is a product model containing engineering knowledge on allowable configurations of modules from a product platform. However, a product platform is not static, but evolves over time with the introduction of new modules. Consequently, engineering knowledge must be maintained in configuration systems. In this paper, it is argued that commercial configuration systems do not sufficiently support knowledge maintenance, since the applied modelling technique does not incite for preservation of the original engineering knowledge. An interaction object is proposed for structuring engineering knowledge on configuration of modules. An example of a cone joint is used to illustrate the modelling power of the proposed interaction object. Finally, it is argued to distinguish behavioural and structural product models in configuration systems, since customers are not in need of a structure, but rather in the behaviour of a structure.

1 Introduction

During the last decades, application of product modelling has gained an increased interest by industry. So far, CAD systems account for the most widespread use of product modelling. CAD systems have evolved from supporting 2D drafting to 3D modelling with feature libraries and integration with analysis and simulation systems, e.g. FEM and mold flow analysis. Other commercial systems for product modelling are now emerging on the market, e.g. PDM and configuration systems. All together, these systems point to a shift of paradigm in product modelling from geometrical modelling to knowledge modelling. For improving the effectiveness and efficiency of the product development process, more knowledge intensive engineering tasks tend to be supported by knowledge-based systems. Research on structuring of engineering knowledge is central for driving this development.

Configuration systems are one of the most promising means to structuring knowledge. Andersson [1] describes configuration systems as expert systems "in which the knowledge of the company 'experts' is captured". A configuration system consists of two subsystems, a modeler system and a configurator system, that supports two distinct processes, a modelling process and a configuration process. This distinction imposes a division of labour between a modeller, i.e. a knowledge engineer, and an end-user, where the knowledge engineer structures knowledge to support the later configuration process performed by the end user. The purpose of the modelling process is to increase the readiness of the configuration process. Therefore, the former modelling process may be considered a work preparation process of the later configuration process. The degree of preparation may range from complete automation of the configuration process [2] to support with a human in control of the process. So far in industry, configuration systems are mostly used for sales configuration, where a sales person can configure a product and prepare a bid on site [3,4]. Thus, configuration systems enable knowledge sharing between the functional areas of a company, e.g. engineering and sales departments.

Modelling in configuration systems may be based upon a systemic viewpoint, where a product is composed by subsystems, also termed modules. The total set of modules constitutes the product platform from which several different products of a product family [5,6] can be composed. Configuration may also be considered from a parametric viewpoint, where the characteristics, also termed attributes, of the modules are variable. By combining the two viewpoints, a module can be modelled as an object with attributes, and the relations between the modules can be modelled by constraints. Thus, the objects and constraints constitute a knowledge base of engineering knowledge ensuring only allowable product to be configured from the product platform.

However, a product platform is not static but evolves over time with the introduction of new modules. Therefore, along with the evolution of a product platform, the knowledge base of a configuration system must be maintained. Where engineering knowledge previously was held partly in documents and partly in the head of the designer, it is now externalized and explicitly structured in a knowledge base in a configuration system. However, companies with years of experience in knowledge structuring in configuration systems report on difficulties in maintaining the knowledge bases. In some cases, knowledge turns out to be as inaccessible in a configuration system as in the head of the designer, since the knowledge engineer is now the bottleneck in accessing the knowledge base. Worst cases report on companies that, due to difficulties in knowledge maintenance, decide to completely rebuild the knowledge base when a product platform is altered. This is a time consuming process for large knowledge bases. Thus, in time, as product platforms evolve in companies utilizing configuration systems, knowledge maintenance will be raised as an important topic in design research on configuration systems.

Several means may be utilized to ease the maintenance of the knowledge base in a configuration system, e.g. keeping the number of constraints small [7] and structuring the constraints in groups [8]. In this paper, another means is pursued, i.e. to preserve the engineering knowledge in the configuration system. The modelling techniques of commercial configuration systems incite transformation of engineering knowledge into simple object and constraint representations. From a configuration viewpoint, this is not a problem, since the original engineering knowledge is not of direct relevance for the end-user, e.g. a sales person. In fact, the knowledge transformation is desirable from the end-users viewpoint, since it lowers the requirements on for example the processing speed and storage capacity of the configuration system. However, the knowledge transformation is a problem from a maintenance viewpoint, since the original engineering knowledge is lost in the transformation, and there is no support for backward reasoning to this knowledge.

The thesis of this paper is that preservation of original engineering knowledge eases the maintenance of the knowledge base in configuration systems. The long-term goal of the project is to develop a method for modelling configurable products, and thereby contributes to the development of the configuration technology [9]. Based upon experiences from consultant work in various companies, this paper presents preliminary results of the project. Therefore, the paper is more of an industrial paper than a research paper. So far, the project is in an initial phase, and the contribution of this paper lies as much in the identified problems as in the proposed solutions. The project is carried out in a Danish center for product modelling. The center is a virtual constellation of researchers, consultants, and industrial partners working with product models. The overall aim of the center is to develop operational methods for the application of product models with the purpose of improving the effectiveness and efficiency of business processes, in particular engineering processes. Within the center, research is conducted on strategic aspects of product modelling, object-oriented methods for development of product models, and organizational aspects of application of product models. More information on the activities of the center can be obtained at the homepage www.produktmodeller.dk.

The structure of this paper is divided into four sections. In the following sections, two aspects of preservation of engineering knowledge are treated, i.e. the relation between modules and the relation between behavioural and structural viewpoints on a product. Finally, section four concludes the paper and outlines future work.

2 Modelling of Knowledge on Module Interactions

Prerequisite for fully exploiting the benefits from application of configuration systems is the development of a product platform from where a set of modules can be configured into various products of a product family. The purpose of a product model in a configuration system is to ensure that the modules of a product can only be configured in a correct manner. For instance, consider the configuration of a car [7]. The engine of the car can be one of the four types 2.0i, 2.1i, 2.0s, or Turbo. Skovgaard [7] shows how the guideline "cars with turbo engine should be ordered with ABS brakes" can be represented with a constraint on the configuration of an engine module with a brake module.

2.1
Preservation of Engineering Knowledge

However, once the constraint is implemented in the product model of a configuration system, it is the only representation of engineering knowledge on how to configure a car with an engine and a brake. It is reasonable to expect that engineering knowledge on correct configurations of engines and brakes was present at the time of the development of the product platform. However, this engineering knowledge is not preserved in the constraint representation, since the constraint does not explicitly explain why a car with a turbo engine should be configured with an ABS brake.

If later a new engine is introduced to the product platform, e.g. a diesel engine, it is impossible to derive from the constraint alone whether or not this kind of engine should be configured with an ABS brake. Similarly, if later a new brake system is introduced to the product family, it is impossible to derive whether or not this kind of brake can be configured with a turbo engine instead of the ABS brake. Based upon guesswork one may reason that since a car can drive faster with a powerful engine it must have better braking capabilities to maintain the same stop length as a car with a less powerful engine. Therefore, since a turbo engine delivers more power than the other engines and an ABS brake has a better braking capability than a normal brake, a car with a turbo engine should be ordered with an ABS brake. However, this reasoning is based upon pure guesswork, and there may be other reasons for the constraint.

Furthermore, if later an existing module in the product family is redesigned, or a new module is developed, one may ask to what extent the knowledge in the product model supports this development. From a structural viewpoint, the whole-part contribution of each module is represented in the product model, whereas from a behavioural viewpoint the contribution of each module to the realization of the behaviour of the product is not represented. Again, in relation to the car example, one may reason that the function of an engine is to transform the energy content of the fuel to mechanical energy. However, again this reasoning is based upon guesswork. Thus, the knowledge content in a configuration system is depreciating the original engineering knowledge.

For more extensive modelling of objects and constraints, attributes can be used in most configuration systems. In relation to the car example, a power attribute for engine objects and a braking capability attribute for brake objects may be introduced. Given these attributes, the constraint may be formally expressed as "engine.power $\leq$ k x brake.braking capability". This constraint preserves more of the original engineering knowledge than the former representation simply stating, "cars with turbo engine should be ordered with ABS brakes". However, the constraint does not preserve the engineering knowledge for determining the constant k.

In most configuration systems, any description in the form of a plain text or graphic files can be associated to objects and constraints. Therefore, one may argue that the original engineering knowledge may be expressed in this manner. However, text and graphics are an informal knowledge representation. If a configuration system is to reason based upon the content of a product model, the knowledge must be represented in a formal manner. Thus, a more formal representation of engineering knowledge is searched for in this paper.

2.2
Structuring of Interaction Knowledge

An essential knowledge component in allowing only correct configurations is knowledge on how modules are allowed to interact when configured in a product. The remaining part of this chapter focuses on engineering knowledge on module interactions, henceforward termed interaction knowledge. Other terms than interaction may be used, e.g. interconnection [10], port [11], or interface [12]. However, these terms imply a physical contact between the modules, which may not be the case. In the next chapter, another kind of engineering knowledge is treated, i.e. knowledge on the relation between behaviour and structure of a configuration.

A central question is where to structure interaction knowledge. At least four solutions exist to this problem, i.e. to structure the knowledge in:

- a separate knowledge base on module interactions, or
- the object of one of the interacting modules, or
- the parent object in a whole-part structure of the interacting modules, or
- an interaction object.

The first solution implies a knowledge base only containing interaction knowledge. For instance, the Baan configuration system has a separate library for constraints [8]. However, this solution enforces a separation of objects from the knowledge on how objects interact. Although constraints may be clustered in groups, the solution does not offer a means for structuring interaction knowledge by directly relating it to the objects representing the modules of relevance.

Another solution is to structure interaction knowledge in the objects representing the modules. However, to avoid redundancy interaction knowledge can only be structured in a single object. Therefore, the objects representing the other modules must have a relation to the object structuring the interaction knowledge. For instance, object-oriented modeling [13, 14] and CRC-cards [15, 16] utilize instance connections and collaborations respectively to model this kind of relation between objects.

Objects can be structured in a whole-part structure representing the assembly structure of the modules. Therefore, instead of structuring interaction knowledge in one of the objects knowledge may also be structured in the whole-object (parent) of the part-objects (child) representing the interacting modules. However, the whole-object may have more part-objects than those relevant for the module interaction. Therefore, interaction knowledge structured in a whole-object requires additional relations to the subset of part-objects that are involved in the interaction.

Finally, another solution is to structure interaction knowledge in a separate object. This object is termed an interaction object, since it models the interaction between two or more modules. In the following, a configuration of a milling machine and a tool is used to illustrate the modeling power of the interaction object.

2.3
An Interaction Object for a Cone Joint

Cone joints are widely used to assemble tools with milling and drilling machines, lathes, etc. A cone joint is a machine element in the category of hub-shaft connections based upon a frictional lock. Besides allowing for assembly/ disassembly, the cone joint is to transfer a torque from the milling machine to the tool. In this example, the reaction force of the cutting operation is neglected. Several norms exist for cone joints, e.g. DIN 228 for tools with metrical or Morse cones.

Consider a product family of six milling machines and four tools. The cone of the milling machine and the tool can be of several types. The number of constraints required to model the configurations of milling machines and tools depends on the number of milling machines, tools, and cone types. In the example shown in Fig. 1, eight constraints, represented with a line between a milling machine and a tool, are needed to model the configuration of three types of cones. Each constraint represents knowledge on the specific configuration of a milling machine and a tool.

Milling machine obects

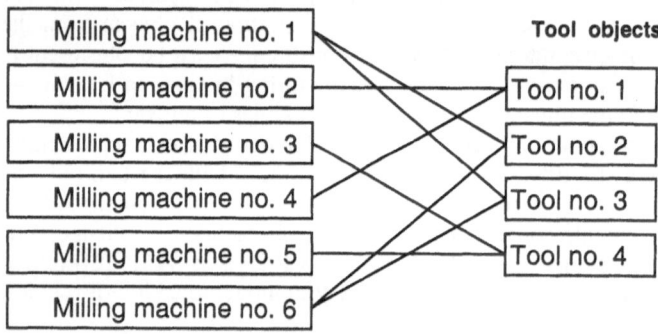

Fig. 1. Configuration constraints between milling machines and tools

Consider now the introduction of a new milling machine to the product family. Consequently, constraints must be reconsidered for the allowable configuration of the new milling machine with each of the tools belonging to the product family. This may be a tedious task for a product family with a huge number of tools. Furthermore, the existing constraints do not explicitly represent general knowledge on the configuration of a milling machine with a tool. Only specific knowledge on each kind of configuration is represented by the constraints. However, from constraints representing specific knowledge one may implicitly reason to general knowledge, e.g. only cone types of similar types can be configured. Nonetheless, such reasoning involves an element of guesswork.

A configuration of modules may instead be modelled by introducing an interaction object as shown in Fig. 2. The interaction object contains generic constraints, e.g. the cone type of the milling machine must be similar to the cone type of the tool. Consider again the introduction of a new milling machine to the product family. By utilizing an interaction object, it is unnecessary to reconsider the configuration of the new milling machine with all the tools in the product family. The interaction object already contains generic knowledge necessary for reasoning to which specific tools the new milling machine can be configured.

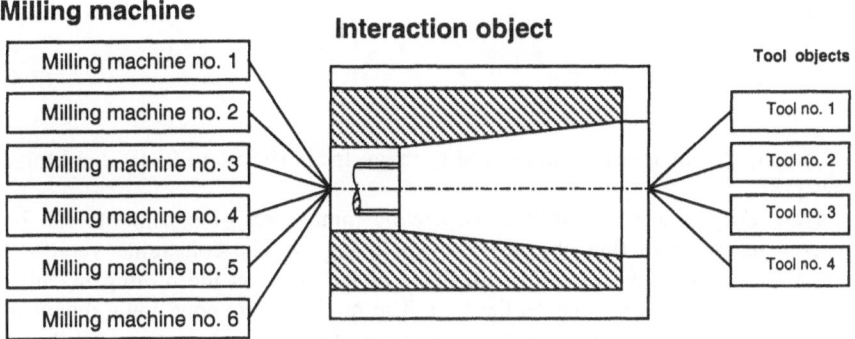

Fig. 2. Interaction object relating milling machine objects and tool objects.

2.4
Structural Configuration of Cone Joint

A cone joint can be characterized by a set of geometrical attributes as shown in Fig. 3. Note the geometry of the cone joint is over-determined, since only three geometrical attributes are required to completely define the geometry of the cone joint.

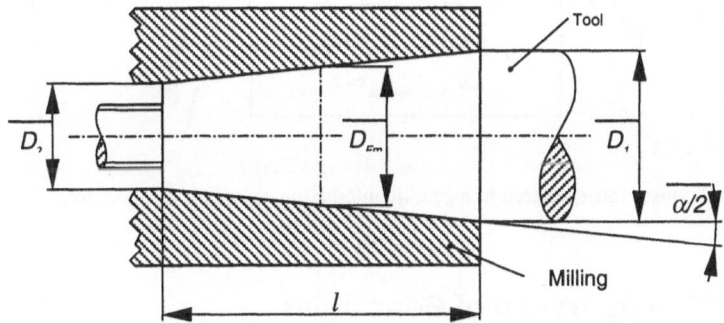

Fig. 3. Geometrical attributes of a cone joint.

The relations between the geometrical attributes of the cone joint can be expressed by equation (1) and (2).

$$\tan\left(\frac{\alpha}{2}\right) = \frac{D_1 - D_2}{2 \times l} \tag{2:1}$$

$$D_{Fm} = \frac{D_1 + D_2}{2} \tag{2:2}$$

Furthermore, according to DIN 254, the cone ratio denoted C is given by equation (3).

$$C = \frac{D_1 - D_2}{l} \tag{2:3}$$

C equals 20 for metrical cones and C range from 19,212 to 20,02 for Morse cones [17].

Given the geometrical attributes of a cone joint, a set of constraints can be clustered in an interaction object as shown in Fig. 4. The constraints represent geometrical relations between the attributes of the cone on the milling machine and the attributes of the cone on the tool. The model of the cone joint shown in Fig. 4 is based upon a parameterised cone joint, where the geometrical attributes are included. Thus, compared to the former interaction object, the parameterised interaction object contains more concrete knowledge on configuration of a cone joint.

Milling machine   Tool objects

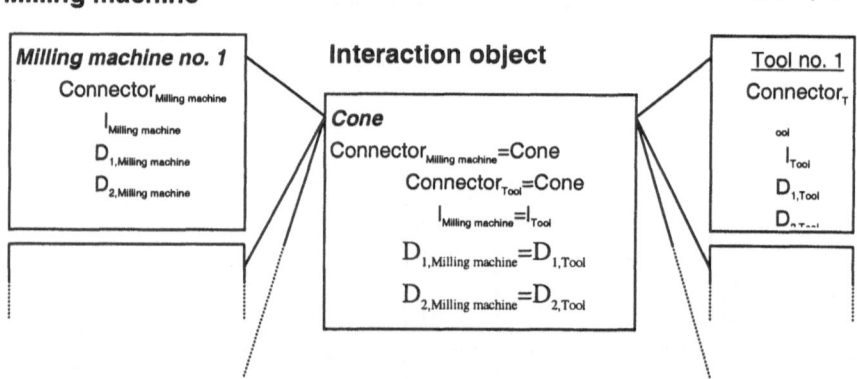

Fig. 4: Geometrical constraints clustered in a generic interaction object for a cone joint.

2.5
Behavioural Configuration of Cone Joint

The constraints on the geometrical relations between the two cones ensure a correct structural configuration, i.e. the fitness of the two cones. However, the constraints do not ensure a proper behaviour of the cone joint, i.e. to transfer a torque from the milling machine to the tool. Therefore, additional constraints are required to represent the behavioural aspect of the configuration. As mentioned earlier, the transfer of torque in a cone joint is based upon a frictional lock. According to [17], the assembly force, denoted with F, required to transfer the torque, denoted with T, is given by equation (4).

$$F = \frac{2 \times c \times T \times \sin\left(\frac{\alpha}{2} + \tan^{-1}(\mu)\right)}{D_{Fm} \times \mu} \qquad (2:4)$$

where μ denotes the frictional coefficient between the materials of the two cones, and c_B denotes a factor considering dynamic operation conditions.

The normal force necessary for creating a friction force causes a pressure on the cone surface. Therefore, the maximum surface pressure of the cone materials limits the maximum torque to be transferred by the cone joint, as expressed in equation (5).

$$T \le \frac{D_{Fm}^2 \times \mu \times \pi \times l}{2 \times c_B \times \cos\left(\frac{\alpha}{2} + \tan^{-1}(\mu)\right)} \times \rho_F \qquad (2:5)$$

where p_F denotes the permitted surface pressure of the weakest material of the cone joint.

Thus, by including equations (1) to (5) in the interaction object of the cone joint, as well as the constraint shown earlier in Fig. 4, a generic interaction object is created for a cone joint. The interaction object preserves the original engineering knowledge on both the behaviour and structure of a cone joint. Therefore, maintenance of the knowledge base is eased, since one may reason without guesswork to correct configurations, when new tools or milling machines is added to the product family. A drawback of the interaction object seems to be that interaction knowledge is more complex to represent. However, if the benefit from an easier maintenance is taken into account, the effort may be worthwhile from a cost/benefit viewpoint.

So far, the interaction object is only conceptualised and only one example is given on its modeling power. The goal is to develop a generic interaction object for representing module interactions in configuration systems. The next step is to include various design theories on module interactions, e.g. connections [18,19], ports in bond graphs [20], and organs in Theory of Technical Systems [21] and Theory of Domains [22]. Furthermore, the modeling technique of commercial configuration systems is going to be exploited for structuring of interaction knowledge.

3 Domain Modeling in Configuration Systems

In the previous section, engineering knowledge on module interaction is treated. However, more kinds of engineering knowledge exist. In this section, another kind of engineering knowledge is treated, i.e. knowledge on the relation between behaviour and structure of a configuration. It is argued that configuration systems besides a structural product model also must include a behavioural model of the product.

3.1
Behavioural and Structural Viewpoints on Configurable Products

According to system theory [21, 22] and cybernetics [23], any object, artificial as well as natural, may be considered from two viewpoints, i.e. from a structural viewpoint of "what the object is", and from a behavioural viewpoint of "what the object does" [24]. For instance, from a structural viewpoint an electrical motor is an assembly of a stator, a rotor, a set of bearings, a housing, etc., whereas from a behavioural viewpoint the electrical motor is converting electrical energy to mechanical energy. Thus, product modeling involves behavioural and structural modeling.

Product modeling in most configuration systems is based upon creating objects and constraints that represent modules and their relations. Product models in configuration systems tend to focus on structural modeling, where the modules are viewed upon as the configurable structural subsystems of the product, i.e. assemblies and parts. For instance, consider the product model of a parabolic antenna used as an example in [25]. A parabolic antenna consists of a set of modules including a reflector, an arm, wall fittings, one more low noise block-converter (LNB, an integrated microwave receiver, amplifier, and frequency converter), LNB holders, a motor, and a receiver box. Each module is a structural subsystem of the parabolic antenna system. Some of the modules are required in a configuration of a parabolic antenna, whereas others are optional. The product model of the parabolic antenna proposed in [25] consists of a set of objects and a set of constraints that supports the structural configuration of parabolic antennas.

However, a customer is not in need of a parabolic antenna from a structural viewpoint, i.e. he is not interested in an assembly of a reflector, an arm and so forth. A customer is in need of receiving satellite channels for a television or a radio, i.e. he is interested in the parabolic antenna from a behavioural viewpoint. Therefore, a customer may only express his needs by the satellite channels of interest to receive. Most often, a customer has more than a single criterion for a product. For a parabolic antenna, other criteria may be price, mounting possibilities on wall or antenna mast, compatibility with other audio/visual equipment, robustness to weather condition, expected lifetime due to corrosion, future possibility to receive other satellite channels, etc. All these criteria are behavioural, and until these attributes are expressed by the customer, a parabolic antenna cannot be configured from a structural viewpoint.

3.2
The Misconception of Structural Configuration in a Sales Situation

One may distinguish various steps in configuration of a product, e.g. sales, product, and production configuration. However, focusing on structural configuration in sales configuration systems is, according to the author, based upon a misconception of the sales situation, or rather the buying situation from the viewpoint of the customer. The following two statements support this argument:

- A customer is not in need of a structure, but in the behaviour of the structure.
- From a mass customization viewpoint [26], a customer is not interested in the configuration of a product, but in buying a product that fulfils his needs.

Clearly, a sales configuration system is a sales person's tool. However, if a sales configuration system is to be a front office system reaching to the customer, one may wonder why there is not a complementary system for the customer. This system is not a configuration system but rather a means by which the customer can express his needs. The sales persons and/or the sales configuration system are then to translate the behavioural needs of the customer to a structural specification of a product that may be configured by modules. This translation has similarities with the relation between customer requirements and technical specifications in the quality function deployment method (QFD) [27]. Since designing is about specifying products fulfilling the needs of the customer, knowledge on translating behavioural needs into a structural specification ought to exist in a company. Otherwise, efforts are better invested in improving the working pattern of the design department than implementing a configuration system. Thus, another kind of knowledge may also be preserved in a configuration system, i.e. knowledge on translating behavioural needs into a structural specification, i.e. the mapping between a behavioural and structural viewpoint on a product.

However, only few of the sales configuration systems implemented in industry known by the author has a behavioural viewpoint on the product. For some reason, that is still a mystery to the author, companies tend to focus only on a structural viewpoint on their products. For instance, consider again the configuration of a car in [7]. Most likely, the size of the engine is only a matter of pride of ownership. More important is the fuel economy, maximum velocity, ability to accelerate, price of service, etc. All these attributes are behavioural, where some are derived from the structural attributes of the engine, while others are influenced by the structural attributes of other modules of a car.

3.3
Domain Mapping Between Customer Needs and Structural Modules

In some cases, a one-to-one causality exists between a need of the customer and the ability of a configured product to fulfil this need if a specific module is included. For instance, consider the configuration of a bicycle. If a customer has a need for carrying luggage on the bicycle, a luggage carrier module is included in the configuration. In this case, there is a direct relation between a behavioural need and the structural inclusion of a module.

In other cases, the fulfilment of a behavioural need cannot be realized by solely including a single module in a configuration. Consider again the parabolic antenna. Given the geographical location of the parabolic antenna, various satellite channels can be received according to the size of the reflector, the number and types of LNB's, the inclusion or exclusion of the motor, and the type of receiver box. In this case, the behavioural need of receiving various satellite channels has wider consequences than simply including or excluding a single module.

In configuration systems no distinction is made on the kind of the attribute of an object, i.e. the attribute may be a structural or behavioural attribute. Therefore, one may argue that configurations systems do not prevent behavioural and structural modelling. However, the object-oriented modelling technique suffers from clustering behavioural and structural attributes in the same domain as well as not making explicit the engineering knowledge relating the two kinds of attributes. Thus, two kinds of product model may be implemented in a configuration system, i.e. a behavioural and a structural model, since separating the attributes in a behavioural and structural domain may ease the maintenance of the knowledge base. A customer expressing his needs creates the behavioural model, and subsequently engineering knowledge is utilized to translate the customer needs to a structural model.

4 Conclusion

This paper addresses the maintenance of a knowledge base in a configuration system. It is argued that maintenance may be eased by preserving the original engineering knowledge, instead of transforming this knowledge into the objects and constraints that are sufficient for sales configuration. An interaction object is proposed for structuring knowledge on configurations of modules, and an example of a cone joint is used to illustrate its modelling power. Furthermore, it is argued that both behavioural and structural product models are required in configuration systems. By these models, engineering knowledge on translating customer needs into a structural specification may also be preserved and, even more important, the sales configuration system puts the customer in focus during the sales/buying process. The project presented in this paper will be furthered by an implementation of a product model in industry with particular focus on maintenance of the knowledge base.

Only a subset of the total engineering knowledge that may be structured in a configuration system is treated in this paper. Future work will cover more kinds of engineering knowledge, e.g. predicting the behaviour of a configuration of objects, and not just the behaviour of the interaction between objects.

Deciding on the degree of preparation on knowledge maintenance involves a cost/benefit analysis. On the one hand, structuring engineering knowledge in the configuration system causes an additional effort in the modelling process. On the other hand, future maintenance of the knowledge base is eased. At this point in the project, there is no answer on an appropriate balance between effort and effect. This paper only proposes an enabling means that may turn out to be an exaggeration of the maintenance problem. More experiences on long-term application of configuration systems need to be gathered before general guidelines on this subject can be derived.

Ever since the development of CAD and MRP systems, the position of the BOM has been discussed, since both systems require a model of the product structure. It is reasonable to expect that a similar discussion will take place on the position of engineering knowledge, since CAD, PDM, and configuration systems may be the owner of this knowledge. This paper only treats structuring of engineering knowledge in configuration systems. However, this is not to conclude that engineering knowledge cannot be structured in CAD or PDM systems. One may claim that only knowledge necessary for configuring a product during sales should be structured in a configuration system, whereas remaining engineering knowledge should be structured in a CAD or PDM system.

Needs vary from customer to customer and thus the importance of the attributes by which the customer may express his needs. In other words, as well as customers require individualized products, they will also require individualized sets of attributes to express their needs. Thus, configuration systems must be able to configure themselves to the customer. This functionality may be seen in the next generation of configuration systems.

5 References

[1]   Andersson, D. M., Agile Product Development for Mass Customization: How to Develop and Deliver Products for Mass Customization, Niche Markets, JIT, Build-to-Order, and Flexible Manufacturing, Irwin Professional Publishing, 1997

[2]   Kota, S., Lee, C.-L., A Knowledge-Based System for Hydraulic Circuit Synthesis, In Engineering Design, Vol. 1, pp. 687-704, 1989

[3]   Esprit 7131, BIDPREP: An Integrated System for Simultaneous Bid Preparation, Tapir Publisher, 1995

[4]   Krömker, M., Thoben, K.-D., Wickner, A., Product Modelling in the Bid Preparation Phase, In Proceedings of ICED 95, Praha, WDK 23, Vol. 4, pp. 1437-1442, 1995

[5]   Meyer, M. H., Lehnerd, A. P., The Power of Product Platforms: Building Value and Cost Leadership, The Free Press, 1997

[6]   Tiihonen, J., Lehtonen, T., Soininen, T., Pulkkinen, A., Sulonen, R., Riitahuhta, A., Modelling Configurable Product Families, In Proceedings of the 4th WDK Workshop on Product Structuring, Delft University of Technology, 1998

[7]   Skovgaard, H. J., salesPLUS a Product Configuration Tool, In Baan Configuration 98.2 Introduction Exercise Guide, pp. 35-43, Baan Front Office Systems, 1999

[8]   BaanConfiguration, Modeler Version 98.2, Baan Front Office Systems, 1998

[9]   Schmitt, E., Manning, H., Paul, Y., Ritter, T.,Tong, J., Configuration in Your Future, In The Forester Report, Forester Research Inc., 1999

[10] Ulrich, K. T., Seering, W. P., Synthesis of Schematic Descriptions in Mechanical Design, In Research in Engineering Design, Vol. 1, No. 1, pp. 3-18, 1989

[11] de Vries, T. J. A., Breunese A. P. J., Structuring product models to facilitate design manipulations, In Proceedings of ICED 95, Praha, WDK 23, Vol. 4, pp. 1430-1436, 1995

[12] O'Grady, P., The Age of Modularity: Using the new world of modular products to revolutionize your corporation, Adams and Steele, 1999

[13] Coad, P., Yourdon, E., Object-Oriented Analysis, Yourdon Press, 1990

[14] Coad, P., Yourdon, E., Object-Oriented Design, Yourdon Press, 1991

[15] Wilkinson, N. M., Using CRC Cards: An Informal Approach to Object-Oriented Development, Cambridge University Press, 1995

[16] Bellin, D., Simone, S. S., The CRC Card Book, Addison-Wesley, 1997

[17] Matek, W., Muhs, D., Witel, H., Roloff/Matek Maschinenelemente: Normungen, Berehnungen, Gestaltungen, Friedr. Vieweg & Sohn, 1987

[18] Roth, K., Die logische Schluss-Matrix, ein Algoritmus zur Analyse und Synthese von Verbindungen und Führungen in der Konstruktion, Fortschritts-Berichte VDI, Reihe 1, Nr. 35, VDI-Verlag, 1974

[19] Roth, K., Methods and relationships for automatic design of connections by the computer, In Engineering Design, Vol. 1, pp. 637-644, 1989

[20] Thoma, J. U., Simulation by Bond Graphs - Introduction to a Graphical Method, Springer-Verlag, 1990

[21] Hubka, V., Eder, W. E., Theory of Technical Systems - A Total Concept for Engineering Design, Springer-Verlag, 1988

[22] Andreasen, M., M., The Theory of Domains, In EDC-Workshop: Understanding Function and Function to Form Evolution, Cambridge, 17-19 June, pp. 21-47, 1992

[23] Klir, J., Valach, M., Cybernetic Modelling, Iliffe Books, 1967

[24] Alexander, C., Notes on the Synthesis of Form, Harvard University Press, 1964

[25] Jørgensen, K. A., Modelling Configurable Products, Department of Production, Aalborg University, 1999

[26] Pine II, B. J., Mass Customization: The New Frontier in Business Competition, Harvard Business School Press, 1993

[27] Hausing, J. R., Clausing, D., The House of Quality, In Harvard Business Review, May-June, pp. 63-73, 1988

Report on Using Product Data Management Software in Managing Modular Products

Markus Vainio-Mattila

Tampere University of Technology
P.O. Box 589
33101 Tampere, Finland

Abstract. This paper describes the execution and results from a survey on the state of product data management in five companies in Tampere region, Finland. All companies have modular products in their strategy hence special attention was paid to that aspect. Since both product data management tools and modular product concept are relatively new concerns, the change process was emphasized in the research.

1 Introduction

Modular product as a concept is not yet well established. However modules refer to a partitioning of the product. A partition can be both physical, existing entity and set of relations in the modelled product data. Different life cycle phases (sales, design, procurement, manufacturing, etc.) and also different disciplines (mechanics, electric, hydraulics, etc.) may have their own way of partitioning the product. In this paper a modular product is partitioned so, that its physical structure in the late phases well correspond to its functional structure of the early phases. By adding or changing modules product's functionality can be changed.

Product Data Management (PDM) is an approach to help companies manage their product structures, product development processes and technical documents. PDM is closely related to enabling software tools and has become more important as the tools have gained efficiency.

A group of Finnish companies in Tampere area and Tampere University of Technology (TUT) have co-operated in "PDM-group" in exchanging PDM knowledge since 1996. A common factor for the companies in the PDM-group is that they all claim to produce modular products. In the beginning of 1999 Tampere University of Technology initiated a project "PDM-evaluation" for the PDM-group to study the upcoming problems and solutions in the PDM implementation projects of five companies. This was a good opportunity to investigate how companies understand and manage modularity in practice.

Not many PDM (or any other Information Technology) tools support modelling of modular products. Thus problems in this area were expected to arise.

In this paper we present the execution and results of PDM evaluation project. In section 2 we set of our research questions and hypothesis and some theoretical background to the assumptions we made. Section 3 is about how we collected the data. In section 4 we describe the industrial cases of this project. In section 5 we present some results and future plans.

2 Theory

In this chapter we present a description PDM of the basic concepts in this report and some problems related to those definitions. We discuss also the subject of this study in more detail.

2.1
Terminology

- *Data* is simple information that can be presented e.g. as value.
- *Knowledge* is more complicated information that can be presented e.g. as logical clauses.
- *Product family* is a set of variants of a product to fulfil a certain set of customer needs. Also know as *product portfolio*.
- *Product architecture* is a concept for how a company understands their product e.g. a formal model that explains the partitioning of the product and the relations between different parts.

Product architecture should also describe how different product individuals are derived from the product family.
- *Unit* is a part in product partition.
- *Function* is a feature of a product wanted by the customer.
- *Interface* is the way a unit effects to its' exterior (not just geometry, but all input and output).
- *Module* is a functional unit with well-defined interface.
- *Non-module* is a unit that has no well-defined functionality or interface.
- *Configuration knowledge* is the knowledge how modules can be combined to achieve valid functioning products.
- *Modular product* is a product family whose variants are achieved by adding, removing, changing, or parametrising modules into non-variant non-modules.
- *Product Data Management (PDM)* is a methodology and a set of tools that help an enterprise to manage both product data and the product development process.

PDM systems keep track of the mass of data and information required to design, manufacture or build, support, and maintain an enterprise's products. PDM integrates and manages processes, applications, and information that define products across multiple systems and media.

By management we understand PDM software functionalities for reading, writing and versioning of the product data according to some specified user dependent process. *In some cases it is rather difficult to make difference between the data and the managing of it.* E.g. a relation between two components of the model expressing that they should not be closer than 10 cm away from each other is clearly of product data, but how about an attribute which shows the last user accessing the documents of those two components?

2.2
Subject of Research

It was stated earlier that not many PDM tools support management of modular products and that all the companies referred in this paper claim to produce modular products. Thus it was quite natural to state our research question as: How is it possible to manage modular products with existing PDM software tools? In following we divide this problem into several sub-questions and hypothesis.

2.2.1
Hypothesis1: Producing Modular Products

Product architecture could be 'integrated' or 'modular':
1. Integrated implying that all components are highly interrelated and thus fixed.
2. Modular so that components form modules which can be changed or added to change the functionality of the product.

Modular product architecture reduces complexity by well-defined interdepencies. Modules with known interfaces can be concurrently developed. Reuse of knowledge and components is possible by developing the product module by module instead of integrated development of the whole product.

On the other hand, modular product architecture allows functional variety by module changes. Modular product architecture can to some extent combine the low cost of mass product and customisability of a one-of-a-kind product.

When defining the modules the cost factors from all life cycle phases should be taken into account:
- Sales / product development
- Creating and maintaining modules and product architecture should not be too arduous. A very wide range of customer needs is hard to meet.
- Design
- Creating and maintaining configure models should not be too arduous.
- Purchasing

Widest possible demand should be met with smallest possible amount of components (the whole product family should be taken into account).
- Manufacturing
  E.g. assembling different modules should be relatively easy.
- Maintenance/upgrading
  E.g. modules should be flexible enough for future upgrading.
- Recycling/disposal
  E.g. dividing biological waste and recyclable plastic into different modules.

2.2.2
Research Question 1: How to Manage Modular Products?

To manage the data involved (relations and entities representing the architecture), Product architecture should be stored, accessed and managed together with the product data in a way that enables the product architecture to be used as a guide for any business process during the whole product life. At least this should be taken into account when planning the data structures of new IT systems, even if all the user interface aspects etc. were not known.

Experience from industry shows that constant refinement of the IT structure from bottom-up, i.e. relying upon existing systems and expanding them is impossible in long term. Most new IT reformations start from top-down, the process of gaining new competence and IT functionality takes so long time that 'first decisions become outdated'. PDM tools seem very natural in initialising any large IT changes and updates.

2.2.3
Hypothesis 2: PDM is a Valid Means for Managing Product Data

A PDM system is used within an enterprise to:
1) Organise, access, and control all data related to its products (product definition data),
2) Manage the lifecycle of that product definition data.
3) Serve as a media during major product and IT changes.

A PDM system may work with a wide variety of software applications (e.g. CAD and VRML environments providing early visualisation), and with traditional non-computer systems that generate product data (such as paper documents). The formal presentation of the product and added documentation properties improve co-working of multiexpertise workforce. Almost all PDM efforts are mind-staggeringly complex and all concretisations are useful. Thus PDM system seems to be the place to store company Product architecture.

The modern day business is subject to short-term engagement of various experts and the PDM system could serve as a meeting place of the workforce, where the knowledge is traded and concentrated efforts occur. Other more ad-hoc solutions, such as improvised integrated databases lead to crucial dependency to some key personnel.

2.2.4
Research Question 2: How to Choose and Implement Software?

To choose and implement adequate software successfully, one should be very well aware of both the data structures the software should be able to process, and the operations it should perform on this data. This is naturally a circular condition; one cannot fix all data structures and operations beforehand. Thus the flexibility and scalability are initial issues.

2.2.5
Hypothesis 3: Modularity Needs to be taken into Account in Choosing PDM Software Tool

Not all PDM software tools support abstract modelling needed to present functions, modules (and their interfaces) and components of the product. Some tools model the product merely as a bill of materials. It was stated in hypothesis 2, that PDM is a valid means for managing product data. However it should be carefully studied, weather the chosen PDM tool supports the company's way of presenting it's product architecture.

2.2.6
Research question 3: What are Modularity needs for PDM implementation?

Chosen software tool should support needed modelling and the pre-evaluation modelling task should be given enough time and resources. At least some examples of product functions, modules, their interfaces and relations to actual parts should be modelled.

Also some study of process modeling should be performed, at least on the relations between modules and organisational entities, and also modules and workflow entities.

In the beginning of the project research question 2 was taken as the subject of this study and we referred to other research in the field to discuss the hypothesis 1-3 and questions 1 and 3. However it was quite clear after the first interview round that project participants did not have a common understanding on the concepts.

Some participants do have functional descriptions to their modules, but interfaces are quite poorly defined. The whole concept of modularity seems to have strong connection to manufacturing and assembly, rather than to design. Therefore the second interview round concentrated in discovering the meaning of both modularity and PDM to the project participants.

3 Data Collection

During May and June 1999, TUT researchers discussed with some PDM project managers and IT personnel in all participating companies. Prior to the meetings questionnaire A was send by e-mail. After the meetings a short description of the situation in each company was circulated. Questionnaire B was sent in July 1999 to explore the modularity aspects in more detail. Second interview round took place during October 1999.

- Questionnaire A had questions in:
- State of PDM (features in use...)
- IT infrastructure
- Users (number, number/project, user grouping...)
- Problems in implementation (technical, organisational, schedule...)
- Some overall questions about modularity
  Questionnaire B had questions in:
- Before starting PDM project, did company have formal description of product structure and process/workflow and the interaction between these two (which tasks concern which parts etc.)?

- What formalism was used in product modeling? (STEP, etc.)
- What formalism was used in process modeling? (PERT, IDEF...)
- If not, are these modeling tasks part of the PDM project?

- Company's concept of modularity
- Does modularity affect organization (design teams assigned module wise)?
- In which activities modularity helps (product development, design, manu-facturing, procurement, sales and after-sales)?
- In which activities PDM helps/is expected to help?
- How does your modular thinking go together with your PDM project?
- Does (chosen) PDM support product structures showing modularity?
- Can modularity be taken into account in workflow management?
- Do you systematically use the workflow and process management features of your PDM system?

4 Description of Cases

As stated before, there are five companies participating the PDM evaluation project. In this chapter we shortly describe the companies by their products, size, locations (Table 1.), IT environment and means of PDM (Table 2.).

Table 1. Companies

	Valmet	Kalmar	Tamglass	Tamrock	Nord berg-Lokomo
Products	Paper, board and fibre machinery and related process control	Solutions for material handling	Safety glass machinery for architectural and automotive glass industry	Machi nery, equip ment and tools for mining and construc tion industries	Mineral and rock processing systems, crushing, grinding, milling and screening equipment and related services, screens, conveyors and other auxiliary equipment
Base locations (all market globally)	Finland, Sweden, UK, Italy, US	Finland, Sweden	Finland, US, Switzerland.	Finland, Sweden, France, Canada	Finland, France, US

Table 2. Results from the first interview round

	Valmet	Kalmar	Tamglass	Tamrock	Nord Berg-Lokomo
State of PDM (features in use...)		Document and production management linked together	Document management	Document, item, authorisation, workflow management	Document, item, authorisation, product structure management
PDM Software	–	DMS + XMAN, piloting MST 9000	Homemade	PTC Windchill	Modultek MST 9000
Docu-ments in database	Planned >1Milion	~10 000 per year	~100000	~100 000	~150 000
No. of design Locations with PDM	Planned 25	1	1	5	2+2
IT – infrastruc-ture CADs used	Catia, AXES, Pro/E, AutoCAD	Catia, VERTEX, AutoCAD	VERTEX	Unigraphics 3D mod.	HELIX, AutoCAD, Catia, Ideas3D
Other than CAD softwares to be linked	BAAN	XMAN	Ms Office	LEAN	ABACUS, POWERE, MFG
Users of PDM No. design/total (all with access to PDM)	1500 / -	70 / 300	- / 130	50 / 200	75 / 350
Number in project development / delivery	20/20	8/4	4/2	10/6	8/4
Active countries	World wide	FIN	FIN	FIN, F	FIN, F, USA
Problems in imple-mentation (technical, organisation al, schedule)				New product (supplier not experienced) importing old data	Distributed access from several pro-duction sites. Product structures easy to change, drawings not

Tamrock, Tamglass and Nordberg-Lokomo start implementation with a pilot project as a local installation and then share the database with other locations. Kalmar tries to do small incremental steps building up an open environment step by step. Valmet prepares for worldwide implementation by thorough evaluation of software vendors/retailers. For software choosing process some feasibility studies were done in each company.

All the companies claim to have modular product strategy, but have quite different idea of the meaning of modularity (Table 3.).

Table 3. Results from the second interview round

	VALMET	KALMAR	TAM GLASS	TAM ROCK	NORD BERG-LOKOMO
Formal description of product structure	Dispersed documents and instruc-tions	Dispersed documents and instructions	Dispersed documents and instructions	Functions and modules partly in PDM	Functional product structure in PDM
Formal isms used in Modeling (STEP, etc. for product) (PERT, IDEF, etc. for process)	Process for ISO 9000?	Process for ISO 9000?	Process for ISO 9000?	PDM internal for product, process for ISO 9000?	PDM internal for product, also process modelling (for ISO 9000?)
Company's concept of modularity	Aims for master model (of product family) for helping the delivery projects	Modules are part of the mechanisms and/or systems, some accessory.	Lot of parametric configuration	Functions connected to com-ponents	Functional modules divided into mechanical, electrical and hydraulics modules
Modularity vs. organization (design teams assigned module wise)?	–	Design teams plan more structurally, designers will specialize more for their own job	No	No	–
	–	Product development and design first and the rest later	Production (parametric), sales (prices), for design interfaces should be defined!	–	Used to be production packages, now functional modules, benefit from sales to after sales
In witch activities PDM helps/is expected to help?	Visualisation, assembly analysis, teams work, important aspect is after sales	–	Document management, patent management	item manage-ment	Sales, design, production and customer views to the product
Does (chosen) PDM support modularity?	–	No	No (not even planned)	Yes	Yes
Workflow managed by PDM?	–	No	No (not even planned)	Yes	No

Problems in the actual implementation were found mostly in communication. Either between the company and a software vendor or between the company and a consulting company, mainly in the phase of customising the software. In some cases adding a direct communication link from the user engineers to the customising software engineer could solve the problem.

5 Results

In this section we discuss the experiences of the participating companies in comparison with the theory and questions presented in section 2. The considerable differences between the companies' products and size etc. imply that they should not aim to completely common means of managing their product families. Tamglass has quite a parametric approach to modularity, using different module width to adjust the width of the manufacturing line. Valmet product is traditionally of one-of-a-kind type, successfully apply modularity to meet enough vide range of customer needs is quite a task. Kalmar has the same kind of problem with customer (freely) defined cockpit layout.

In all the participating companies the functional aspect of module is better understood, studied and documented than the interface aspect. Thus knowledge re-use is still dependent on designers' experience and complexity is not much reduced. Nordberg has gone quite far in adapting its organisation and business processes according to the product architecture. Also Tamrock is heading this direction.

The product architecture is not static. Paper or other stand-alone models go out of date sooner or later. They might be of use in planning the company's strategies, but in operational use dynamic product architecture should be stored in dynamic way to company's IT system. Since PDM tools well support dynamic linking to other software, they seem to be a valid means for this.

Nordberg and Tamrock have functional modules defined in PDM system, configuration knowledge is either stored in sales configurators or not stored at all.

5.1
Specifying Data and Data Management Process

In the process of choosing and implementing PDM system the needed/wanted aspects of the company product family and business process was modelled to form a clear description: 1. What data (concepts used/needed) PDM should manage? 2. What does this managing mean (what user should be able to do with the model in viewing, updating, creating and deleting data)? This was followed by the iterative process with software vendors about what can be done and how much it costs.

Most companies had been using outside consultants to form their specifications. Nevertheless only Nordberg had some formal model (on document version management). Also it seems that evaluating the company need (forming the specification) and evaluating the existing PDM tools was not done separately. Thus the specification was done in the modelling language of the software tool instead of an independent language more familiar to the company.

New PDM systems allow relatively free data structure of entities and the relations between them. This structure should be well defined with examples like in Fig. 1. On the other hand customised functionality (in database queries or user interface) might be expensive.

In Fig. 1. there are shown some functions, modules, components and their relations as an example of data structure. An example of functionality is that a designer glancing at module *power steering* and *small engine* should be presented the interface between these modules through the dependencies between some of their components. These kinds of specifications were missing in all companies.

One question is if the PDM system should also support analysis of the modelled data. Filtered searches to the database according to entity name or attribute value are basic features, but classifying and grouping the data according to multiple relations is more difficult. These kinds of analysis come into question in defining/re-defining the modules. Both Nordberg and Tamrock are implementing classifications and viewing the data from different points of view through simple filtering.

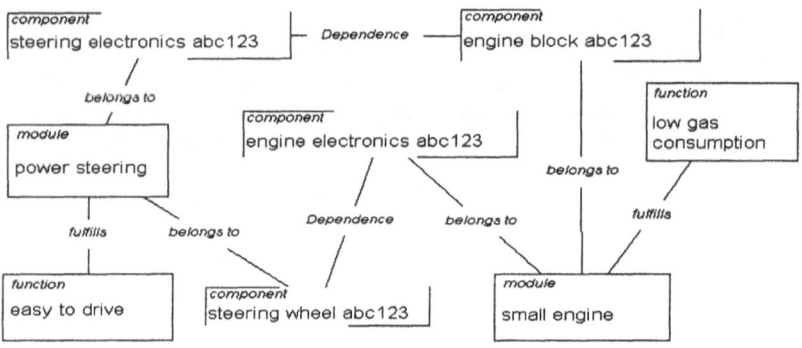

Fig.1. An example of some data entities and relation

An aspect of big product modelling projects is the need to manage incomplete information. Depending on the personnel available to modelling activities, it might take 1-3 years to get a usable models built. During this time the PDM tool should be in everyday use already. Early presentation and visualization of data, however premature, is necessary. In changing to new system, Tamrock started creating new items as early as possible, even though the old data was partly non-valid.

The process of defining the modules can be viewed from components to functions (functions, interfaces and modules are found from an existing component structure), or as functional structures as the starting point. It is quite easy to imagine that both (bottom up and top down) approach are valid in respective cases. Changes to the architecture could occur from either changing a customer requirement/function or an individual component (e.g. due to technology improvement). In both cases module structure with valid interfaces should be maintained.

5.2
Trade Off of Modeling

Some estimation on the costs and benefits of each object, relationship and operation described by the metamodel, i.e. the language used should be done. Also should be estimated how much to go into details, and into which details, in the actual modeling work.

- Is the entity needed/is it used in company processes (how often it is needed)?
- Where and how formal is the entity described now (paper/electric, formal/informal, does the modelling work need to be done from scratch)?
- Are there currently significant problems in the area described by the entity?

The long-span nature of PDM projects is well-observed fact. Increase in resources can shorten the time needed for mature decisions in the modeling work, but only to some extent. Also the ongoing modeling work easily causes iterative changes to the modelled phenomenon itself. The earlier the models become useful to the organisation, the more likely the effort will reach the full scale.

6 Conclusion

In all the participating companies the functional aspect of module is better understood, studied and documented than the interface aspect. Thus knowledge re-use is dependent on designers' experience and product complexity is not much reduced. Two companies have their product architecture stored in PDM to some extent configuration knowledge is either stored in sales configurators or not stored at all.

In the process of choosing and implementing PDM system formal specifications were missing. Also the specification was done in the modelling languages of the software tools instead of an independent language more familiar to the company. Perhaps due to insufficient specification, classifications and viewing the data from different points of view was implemented only through simple filtering.

No estimation on the costs and benefits of each object, relationship and operation described by the metamodel, i.e. the language used was done. Also no estimation on how much to go into details, and into which details, in the actual modeling work.

7 References

[1] CIM data: PDM Buyers Guide 7th Edition, volume 1.
[2] Boynton, A.C., Victor, B., Pine II, B.J. *New competitive strategies: challenges to organizations and information technology*. In IBM Systems Journal, Vol 32, N 1, 1993. pp. 40-64

[3] Chun-Che, H., Kusiak, A.: *Modularity in Design of Products and Systems.* 6th Industrial Engineering Research Conference Proceedings, Miami Beach, FL, May 17-18, 1997.

[4] Ulrich, K., Tung, K: Fundamentals of Product Modularity. Vol. 39, Issues in Design Manufacture/Integration ASME 1991.

[5] Miller, Thomas D.; Elgaard, Per: "Defining Modules, Modularity, and Modularization", Design for Integration in Manufacturing. Proceedings of the 13th IPS Research Seminar, Fuglsoe 1998. ISBN 87-89867-60-2. Aalborg University 1998.

[6] Elgaard, Per; Miller, Thomas D.: "Designing Product Families", Design for Integration in Manufacturing. Proceedings of the 13th IPS Research Seminar, Fuglsoe 1998. ISBN 87-89867-60-2. Aalborg University 1998.

ValueMap™ – a Method for Understanding the Economical Potential of Product Modularisation and Cost of Variety

Tobias Larsson and Johan Åslund

Modular Management AB
P.O.Box 7195
103 88 Stockholm, Sweden
e-mails: tobias.larsson@modular-management.se; johan.aslund@modular-management.se

Abstract. This paper presents a method called the VALUEMAP™. The method provides a means of calculating THE ECONOMICAL POTENTIAL OF PRODUCT MODULARIZATION and an understanding of PRODUCT VARIETY COST. The method also provides a clear graphic representation of the company, which serves as a common foundation for COMMUNICATION between different departments, for instance R&D and marketing, which often have trouble communicating. In the method, costs driven by the product structure come into focus. Key figures such as cost per part number, cost per variant and cost per product are generated during the process.

From the method, derivative tools are developed for both management and design engineers, aiming to COMPARE INDIRECT AND DIRECT COSTS. Management is helped in investment decisions, restructuring the organization, future company cost estimates, corporate value estimates and correction of present product costing methods. Design engineers are supported in concept evaluations, when comparing different technical solutions.

1 Introduction

After several years of consulting in product modularization, Modular Management AB realized that in order for modularization to have real impact in a company a large part of the product range must be modularized. Such a development effort often results in increased capital investment, compared to the company's projects in the past. Development resources, manufacturing tooling and production equipment are some of the areas, which initially need large investments. To be able to argue for such big investments, it is crucial to speak the right language to the right people. The arguments differ between departments and levels in the hierarchy, as shown in Fig. 1.

Investments must pay for themselves, which means big problems for the modularization project leader. He or she must often argue for longer payback periods than usual, because of the extent of the investment. A successful modularization means an annual increase in profit over many years, but may need a longer period to make it happen [Erixon et al 96]. Or does a modularization project really need more time to pay back the initial investment?

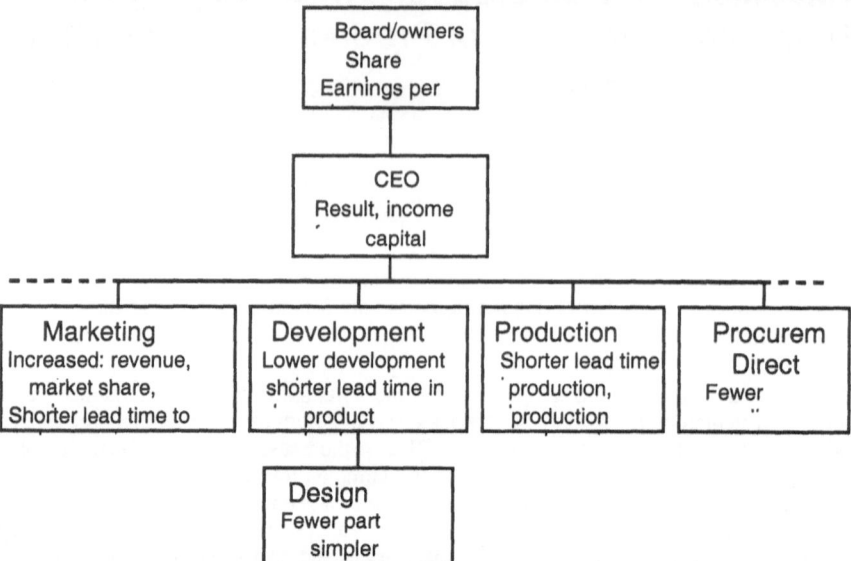

Fig.1. Different levels in the company have different arguments to implement modularisation.

In traditional product costing, typical posts include material, labour and some tooling/machining investments. It may be the case that a modularization project needs a longer payback period due to higher material cost than a traditional development project. But modularization project affects a broader range of costing aspects than a traditional costing calculation tells us. As Fig 2 shows, traditional costing methods often estimate to high costs for a modular product.

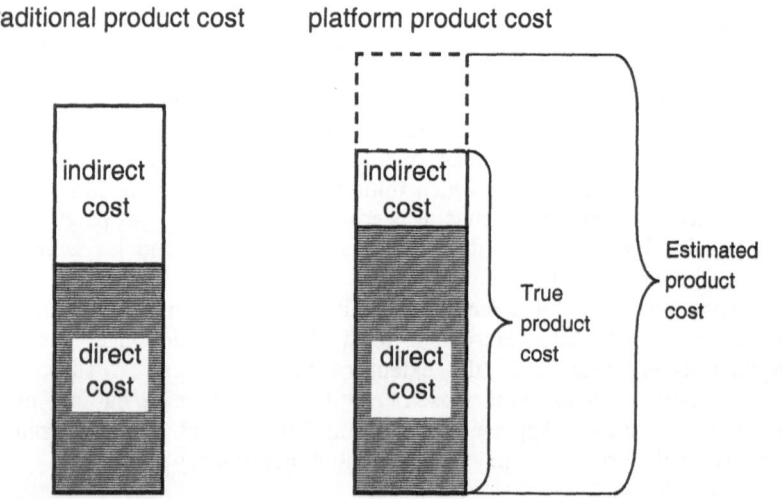

Fig. 2. Traditional product costing will estimate a higher cost for the modular product due to the higher direct costs, even though this is not necessarily true.

This means that we need to better understand what modularization actually does to the company, in monetary terms. Our way to communicate how modularization affects a company, is shown in a ValueMap™ [Larsson, Åslund 98], see Fig 3. The ValueMap™ illustrates how overhead costs are distributed within a company and what they are driven by, for instance by the product structure or by the production volume. From the ValueMap™ several tools are derived, both for management and design engineers, all aiming to guide the company towards the utilization of the modularization.

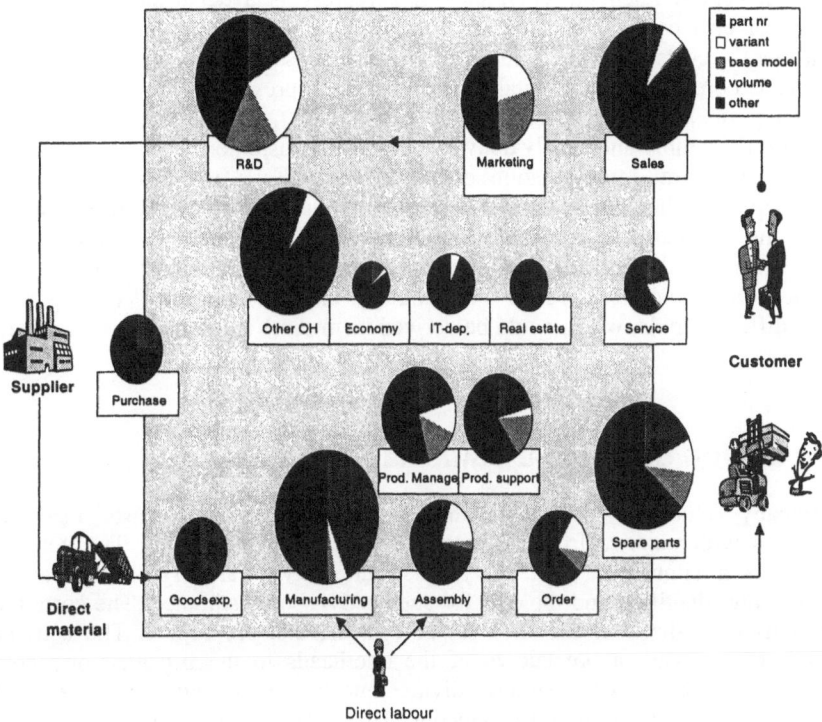

Fig. 3. The ValueMap™ is a graphical representation of a company activity analysis. (This company will be called A in the rest of the paper.)

The ValueMap™ helps management in estimating the economical potential of a modularization. Design engineers are helped when comparing different conceptual designs.

2 The ValueMap™-Description and Examples

This chapter will discuss what a ValueMap™ actually is, how a study is conducted and finally what results have been generated through several ValueMap™ analysis.

2.1
What is a ValueMap™?

The analysis behind a ValueMap™ adapts an Activity Based Costing (ABC)-approach, which involves many interviews, which defines a great number of activities. To enable a presentation of the analysis on one single page, the ValueMap™ was developed. The model graphically illustrates where costs accrue, the relative weight between cost centers, and the driving forces behind the costs at each cost center. Each pie chart in Fig. 3 represents a department (usually a number of cost centers) and the area of the pie chart is a proportional to the department's costs. In the ValueMap™ the process from customer order to delivery is considered as the vital process. A line represents the process with its beginning in a customer order, via sales, R&D, procurement, suppliers, production/assembly and finally delivery back to the customer.

Being a neutral representation of the cost structure of the company, the ValueMap™ can be used as the smallest common denominator in communication within the company. It has been our experience that people from different departments easily makes an agreement using a ValueMap™ in the argumentation. Once realized the inner strength in the ValueMap™ we began to sketch different tools, which could be derived from the ValueMap™, as described in section 3.

2.2
How do you generate a ValueMap™?

A company ValueMap™ is drawn with an activity analysis that consists of several interviews with persons from different parts of the organization. The activity analysis is a simplified ABC-analysis. In an activity analysis a company's activities are identified and the cost for each activity are estimated. The costs for each activity is distributed over a set of cost drivers, see Fig. 4. The activity analysis begins with a breakdown of the overheads to the organization's cost centers. Each cost center's overhead divides into indirect labours, machines, real estate, indirect material and miscellaneous. Thereafter, main activities at the cost center are identified. The activities connect to cost drivers, shown in Fig.4 as number of part numbers, number of product variants, number of base models, production volume and miscellaneous (none of the above). The selection of cost drivers focus on modeling the dependence between indirect cost and the company's product structure. Due to the way the structure is built, the information system in question etc the cost drivers are somewhat customized to each studied company. A typical ValueMap™-study contains about 35 interviews and 250 analyzed activities.

General cost drivers in a ValueMap™ are divided into three categories, product structure dependent, production volume dependent and miscellaneous drivers, which are not driven by the product structure or by the production volume. The product structure dependent drivers are divided into (number of) part numbers, (number of) product variants, (number of) subsystems, (number of) base models or what ever will suit the studied company.

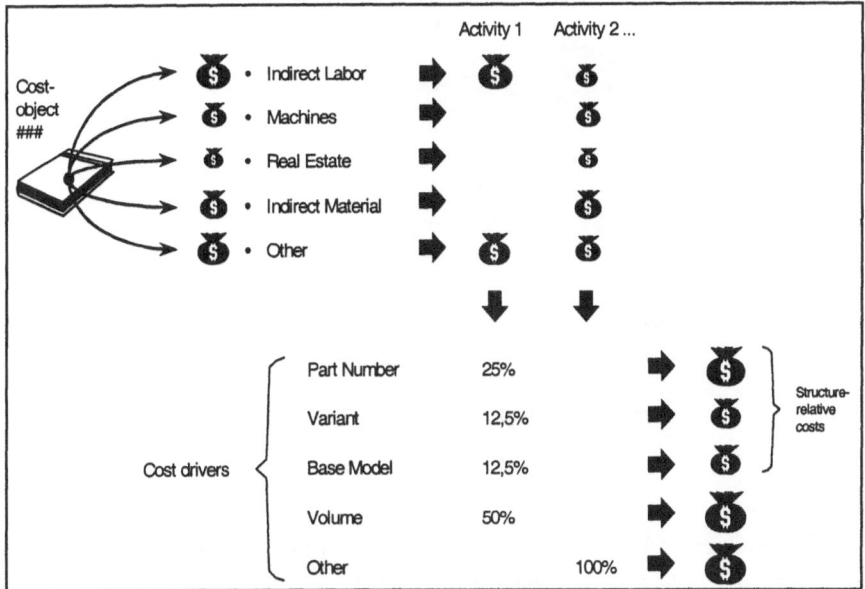

Fig. 4. An illustration of the activity analysis

A modularization renders the possibility to cut product structure costs, often referred to as complexity costs. Note that departments located "close to the product" in Fig. 3 that is close to the order-to-delivery flow have a greater relation to the product structure, e.g. part numbers, product variants and base models, than departments not located on the line.

2.3
Results from ValueMap™-analysis

When conducting the ValueMap™-analysis on several companies we have seen a pattern for the share of product structure dependence. All studied companies have product structure dependent costs of 30-40 % of studied costs, including overhead costs and blue-collar workers, which other studies also supports [SAM 93]. Production volume or "miscellaneous" drive the rest of the costs. "Miscellaneous" is defined as being activity drivers, which have no connection to the product structure or the production volume.

A general conclusion of these studies is that between 30-40% of company cost can be affected by a change in the product structure. Due to the ValueMap™-analysis we can relate a bigger portion of the total cost to the product family design, not just cost for purchased parts and manufacturing and assembly cost. The ValueMap™ shows that many other departments will be affected by a modularization that simplifies the product structure. The immediate implication is better payback and more accurate product cost estimates, see more in chapter 3

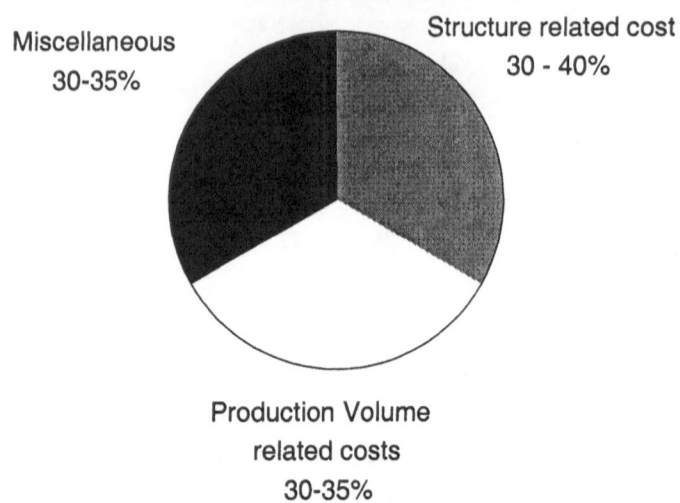

Miscellaneous
30-35%

Structure related cost
30 - 40%

Production Volume
related costs
30-35%

Fig. 5. General results from ValueMap™ studies shows that product structure related costs
are 30-40% of the total costs, excluding direct material.

Imagine a modularization project aiming to redesign most of the product range.
This will affect the whole company, the whole ValueMap™. Furthermore,
imagine that the modularization project results in a potential to reduce 50% of the
product structure complexity, still delivering the same amount (or more) of
product variants to the customers. This will lead to a large saving in costs related
to the product structure.

An illustration of the kind of input a ValueMap™-study gives will be
exemplified by Company A, presented in Fig. 3. Company A had generated a
modularization concept and they wanted to know the economical potential of the
concept. After a ValueMap™-study and a product structure analysis we concluded
that the modularization concept would reduce the product complexity with 50%,
and therefore the product complexity costs with 50%. Company A's module
concept would still deliver the same volume and amount of product variants to the
customers. The factory costs would be reduced with 25%, excluding direct
material. Lead-time in factory reduced with 70% generated reduction in working
capital with 12,5% of the total costs. Company A's development department, for
example, could be reduced to 50% once the modularised product platform had
been introduced to the market.

In projects where we have been able to calculate this type of potential we have
seen a cost reduction potential between 10-25% of the overall plant costs,
excluding direct material. One unique feature with ValueMap™ is that it helps
you to realize the potential, see more in section 3.

3 Conclusions

3.1
Pro and Con with the ValueMap™

Up till now, the ValueMap™-analysis has only included a plant's costs, excluding costs for material. The economical potential for modularisation has been between 10-25% in each case. But, there are even more costs linked to a company that hasn't been studied. Primarily we are talking about costs for material and the supply chain, including sales offices. Fig. 6 illustrates an authentic relation between factory costs, cost for purchased material and supply chain.

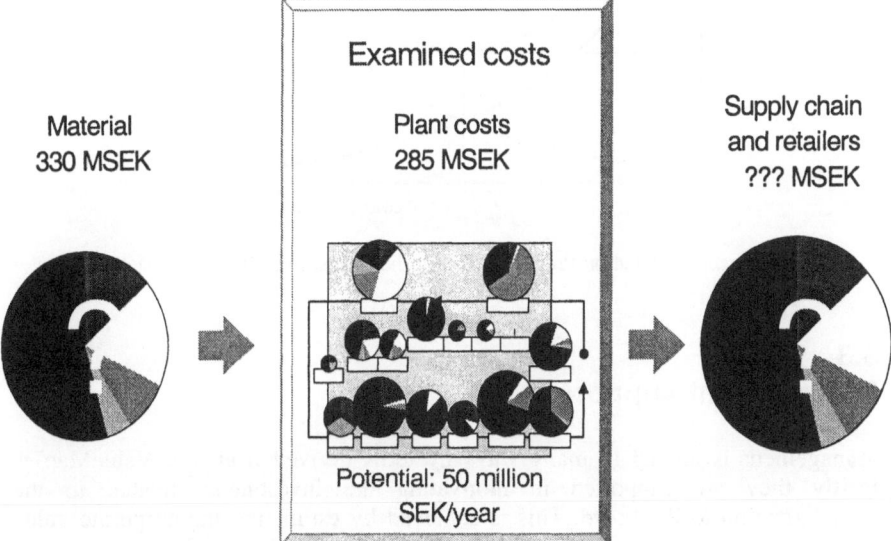

Fig. 6. In addition to the conducted ValueMap™-studies, costs for material, supply chain and sales offices must be added. How is the relation between different activity drivers for purchased material, supply chain and sales offices?

Unlike material and to some extent labour cost the potential of modularisation will not easily put in to practice. A reduction of the product structure related costs would only be achieved through actions. The potential can be attained either through cut-backs on personnel, machinery and real estate, or through reallocation of resources to more value adding activities, for instance new product development, ramp-up etc.

A great benefit with the ValueMap™ is all tools derived from it, which supports management, engineering and marketing in their work, which will be described below.

3.2
Concept Evaluation Process

This tool helps design engineers every time concepts shall be evaluated. The focus of the tool is to compare direct and indirect costs, in order to find the lowest over all costs, view Fig. 7. The tool has four corner pillars: cost, customer value, internal demands and risk [Larsson, Åslund 98].

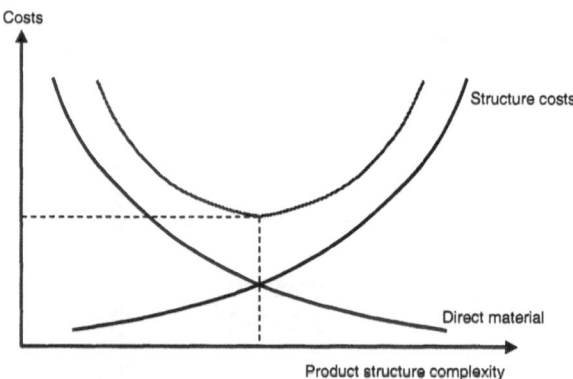

Fig. 7. The Concept Evaluation Process looks for the lowest overall cost for each concept.

3.3
Management support

Management is helped in many ways by tools derived from the ValueMap™. Firstly they are supported in motivating the investments related to the modularization to the board. This is achieved by estimating the corporate value after modularization. Secondly the product-costing model can be adjusted to a modularized product structure, using the ValueMap™-analysis as an aid. Thirdly the ValueMap™ helps when the potential of modularization shall be realized.

4 References

[1] [SAM] Samarbetande konsulter Lönsamma kunder lönsamma företag; ABC-teknikens grunder (Profitable customers profitable companies; the basis of ABC.) Brombergs förlag. 1997 (1993) pp 40-43 [in swedish]
[2] Larsson, T., Åslund, J. Process for economic evaluation of design concepts. Royal Institute of Technology. Stockholm, Sweden. ISSN 1403-7777. Master thesis (1998) [in swedish]
[3] Erixon G., Fredriksson J., Romson L., and von Yxkull A., Modulindelning I Praktiken, Sveriges Verkstadsindustrier (in Swedish only), ISBN 91-7548-460-9, 1996. pp 135